Problems in the Behavioural Sciences

Hunger

Problems in the Behavioural Sciences

1 Contemporary animal learning theory
A. DICKINSON

2 Thirst
B. J. ROLLS & E. T. ROLLS

3 Hunger
J. LE MAGNEN

4 Motivational systems
F. TOATES

Hunger

J. Le Magnen
Laboratoire de Neurophysiologie Sensorielle et Comportementale,
Collège de France,
Paris, France

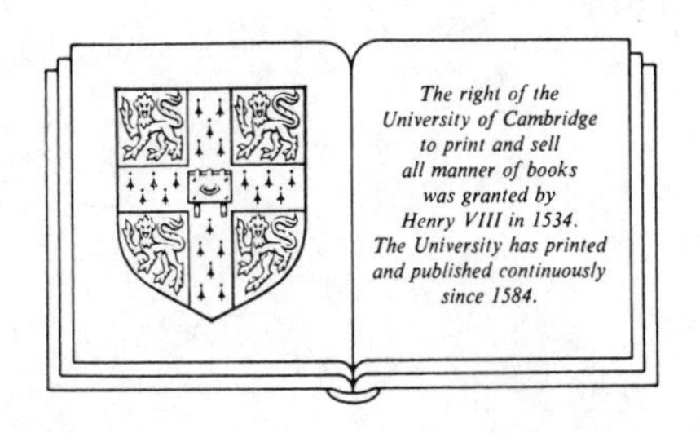

CAMBRIDGE UNIVERSITY PRESS
Cambridge
London New York New Rochelle
Melbourne Sydney

Published by the Press Syndicate of the University of Cambridge
The Pitt Building, Trumpington Street, Cambridge CB2 1RP
32 East 57th Street, New York, NY 10022, USA
10 Stamford Road, Oakleigh, Melbourne 3166, Australia

First published 1985

Printed in Great Britain by the University Press, Cambridge

British Library cataloguing in publication data

Le Magnen, J.
Hunger.—(Problems in the behavioural sciences)
1. Hunger
I. Title II. Series
591.1′3 QP138

Library of Congress cataloguing in publication data

Le Magnen, Jacques.
Hunger.

(Problems in the behavioural sciences)
Bibliography: p.
Includes index.
1. Hunger—Physiological aspects. 2. Hunger—Psychological aspects. 3. Brain—Localization of functions. I. Title. II. Series.
QP138.L4 1985 599′.013 85-7729

ISBN 0 521 26450 2 hard covers
ISBN 0 521 31122 5 paperback

UP

Contents

Foreword

Hunger (like its predecessor in this series, *Thirst*, by B. J. Rolls and E. T. Rolls) is inter-disciplinary in its very nature: an adequate approach to the study of food intake, and the mechanisms that subserve it, is possible only if one starts by ignoring the traditional boundaries between psychology and physiology. Professor Jacques Le Magnen shows elegantly how, if one first poses essentially behavioural questions and then, to answer them, makes a series of simple (but careful and well-controlled) behavioural measurements, one is inevitably drawn to consider the physiological mechanisms that are likely to underlie the behavioural regularities observed; also, the physiological questions that one can then go on to pose are much more precisely framed and better targetted than if one had simply gone into the endocrine and nervous systems and baldly asked, 'how does this tissue control food intake?' There is indeed no better example of the importance of the systematic description of behaviour for problems in physiology than Le Magnen's now-famous observation that meal size is well correlated with the latency to eat the next meal, but not with the interval since the previous one. How much can flow from such a seemingly simple observation the reader will find (with much else) here.

Professor Le Magnen's laboratory is located in Paris, in an institution at once *ancien régime* and entirely modern, the Collège de France. It will surprise no one that our guide to the mysteries of hunger should reside in the city in which a million master chefs have explored every nuance of the motto, 'l'appétit vient en mangeant'. A future generation of chefs may perhaps have to pass an examination on the relation between this essential gastronomic truth and fluctuations (here lovingly described) in the levels of insulin and other hormones that regulate the concentration of sugar in the blood. They will find all they need for their studies in this book. Who knows, their labours may even enable them to add fresh tricks to their trade and to shape the next 'nouvelle cuisine'. For Professor Le Magnen shows the keen interest of the Frenchman, as well as the nutritional scientist, in all the varied aspects of that most important part of life, eating. His lucid account of how they can nearly all be encompassed by a few simple principles of wide generality makes compelling reading.

Acknowledgements

Solange Fanjat de Saint Font is thanked for her important help in the preparation of the manuscript and for her work in collecting references.

Jeanine Louis-Sylvestre, Christiane Larue-Achagiotis, France Bellisle and Michel Devos, other coworkers of the author, are thanked for their work which is cited and illustrated by them in this book.

Introduction

This book entitled *Hunger* deals with experimental studies of food intake and of the latter's role in nutritional homeostasis. Historically, from Aristotle until the first decades of this century, these studies have been limited and obscured by the fact that hunger was viewed only as a subjectively experienced feeling. Despite trivial evidence for relationships between this feeling and food deprivation, and for the relief of hunger by food intake, speculations and investigations have long focussed on the perceptual components of the hunger sensation. It was thought that such components, together with where they originate, were causes of both the hunger sensation and an induced state of the central nervous system associated with the acceptance and intake of foods. The notions of need, drive or motivation that create this state were unable to provide a mechanistic approach to solving the problem.

Only recently, rather than merely evaluating the intensity of hunger in man, investigators have begun to measure food intake. As in other fields of the behavioural neurosciences, this transition from subjective to objective studies was primarily the result of the development of experimental procedures and of techniques of measurements which had been applied to animal models. Such animal studies address the questions of how a combination of internal and external signals governs the selection and intake of foods, and how feeding behaviour is incorporated into the overall process of nutritional homeostasis. In the last few decades, experimental data have provided answers to these questions which go beyond the speculative theories of the past. This basic knowledge obtained from animal studies has permitted research to return, with a solid background, to humans. The results of their main conclusions will be reviewed in this short book.

It is self-evident that nutrition is the main requirement of all living systems. The first step and prerequisite of nutrition, and therefore of growth, self-maintenance and reproduction, is the selection from the environment and subsequent intake of substances designated 'nutrients'. Active feeding, that is to say feeding preceded by food-seeking behaviour, is a necessity of animal life. It is the main characteristic highlighting the difference between the animal and vegetal kingdoms. In the former, high selective pressure results from the essential survival value of feeding activity. It was and still is the most powerful agent in the evolution of species. Through natural selection, it was the origin not only of differentiation of organs devoted to eating but also of a somesthetic nervous system.

That natural selection is dependent on behaviour is often overlooked. Before mating, in order to be able to transmit their genetic heritage, animals have to survive until maturity thanks to a series of regulatory behaviours – among these the efficient feeding behaviour of themselves and their young. The best fit sensory and motor systems, and brain processing of neural input and output, have been selected and fixed through evolution. The least defect of these systems and the resultant failure of proper feeding eliminated individuals from being able to reproduce and could have led to the disappearance of whole species.

Man has emerged from this animal history presumably because he has developed the best capacities required to solve his problems of food sources and self-defence. The most important step in human evolution has been an extension of food seeking by the invention of the production of foods: cultivation and breeding. Presumably, there was the origin of social organization. Today there are still some societies in the world subject to the cruel laws of natural selection which operated in the first ages of humanity.

Out of this specific problem of 'hunger in the world', feeding behaviour – because it is the main daily condition of the maintenance of life – is highly socialized and a source of socialization. Many things could be and have been said by sociologists and anthropologists about the social role of the meal. The sociocultural aspect of food selection and food likes and dislikes superimposed on purely physiological determinants are of the greatest importance in human feeding behaviour. This makes it difficult for us to extrapolate results from animal studies. However, the role of sociocultural factors as external reinforcers of human food habits will often be stressed in this book.

The attribute of the human dimension is to develop a science, ethics and aesthetics. Food is an object of science; but also possesses a symbolic meaning. It is the basis for many rites in all religions. The offering of food belongs to ethical customs in many cultures. In addition to food science and ethics, man has made the preparation of foods an art, and feeding behaviour an aesthetic activity. Beyond, and often in opposition to, the physiological role of food and feeding (which will be treated in this book), preparing and tasting the *Grande Cuisine* is one of the most refined human behaviours.

1 Experimental techniques and procedures

A quantitative assessment and experimental study of feeding behaviour has been made possible thanks to the development of appropriate techniques and procedures. The latter have generally been associated with the simultaneous control or manipulation of variables which affect the intake of food or are consequences of this intake.

Assessment of food intake and parallel measurements in unaltered conditions

Animal models

The animal model which is used most often is the rat. Specific studies have been carried out with other rodents (guinea pig, hamster, gerbil, rabbit) and mammals (dog, pig, sheep and cattle). A limited number of investigations have been made with monkeys and particular mention must be made of experiments using hibernators such as deer-mouse and ground squirrel. Comparisons have been made between various strains of rat, often in association with genetic investigations. Comparison of lean and obese littermates in heterozygous strains of rats and mice which are genetically obese (Zucker *fa/fa* rats and *ob/ob* mice) has been a fruitful tool in the investigation of the relation between food intake and body energy balance.

Food intake measurements in animal models

Whatever other variables may be controlled or manipulated at the same time, the most important measurement is that of the intake of food itself. Surprisingly, until recently the short- or long-term measurement of the amount eaten was made using techniques that were far from perfect

Feeding in animals (as will be more extensively discussed in Chapter 2) is essentially periodic in nature. Rats, like other animal species and like human beings, are not nibblers but are meal eaters. Their feeding habit, when allowed free access to a permanently available food source, is a succession of bouts of eating (or 'meals') with meal-to-meal intervals of satiety. Another condition of feeding, and of its measurement, is the feeding schedule: experimental animals are presented with food once or several times daily for periods of one or more hours. Finally, the feeding response studied and to be measured can be limited to one meal, as is the case in many studies on determinants of meal size.

The simplest technique for measurement, still commonly used, is the manual weighing of what remains in the food-cup at the end of a given period of observation. Spillage can be avoided or controlled by the use of food-cups with special openings. This simple technique is adequate when the feeding response to be measured is limited to a particular meal, but it is not adequate in medium- or long-term feeding *ad libitum* (free feeding).

Two methods have been developed which permit the automatic and quantitative recording of moment-to-moment feeding behaviour of rats and other rodents. The more long-standing is the recording of lever-pressing for food pellets (Skinner, 1930; Anliker & Mayer, 1956). Rats are trained to press a lever to obtain a specific amount of pelleted solid food (50 mg). Though better than the simple weighing of the food remains, this has nevertheless some defects and limitations. The rat has to be trained initially. The requirement to press a lever to obtain an arbitrarily chosen amount of food is an artificial condition which modifies the spontaneous microstructure of the eating behaviour. A comparison of feeding patterns recorded in the same rats by this and another technique (described below) shows differences in sizes, durations and rate of meal consumption recorded before and after elimination of lever-pressing (Kissileff, 1970). In addition to the crumbling of the meal brought about by delivery in pellet form, the technique excludes the sensory (visual and olfactory) action of the offered food in initiation and maintenance of the eating response. Finally, the lever-pressing technique excludes the use of various diets or regimens such as greasy food, powdered sucrose, and so on, which cannot be produced as pellets.

Eliminating the lever has been an improvement. In an apparatus called the 'eatometer', a rat or mouse makes the pellet fall by putting its head in a hole and, in so doing, interrupts an infrared beam (Kissileff, 1970; Wiepkema, 1971); these interruptions are recorded.

The best and most recent means of measuring food intake is continuous automatic graphic recording of the weight of food-cups. The original automatic electro-mechanical balance of Pokrovsky & Le Magnen (1963) has now been replaced by more compact devices. The food-cup, loaded daily with 50 g of food, is continuously weighed by a strain-gauge. The signal, from 0 to 10 mV, is proportional to the weight of the food-cup, and therefore to the amount of food eaten; it is graphically recorded to an accuracy of 0.1 g and a time resolution of some seconds. A set of 10 to 12 food-cup-weighing devices, each of which has been placed in an individual cage, is connected to a poly-channel graphic recorder. In addition, signals may be recorded and processed by computer (Geiselman *et al.*, 1979; Strohmayer *et al.*, 1980).

Such recordings of feeding pattern which allow meal consumption and meal-to-meal intervals of satiety to be analysed raised the question of how to establish criteria for identifying bouts of eating as 'meals'. Rats eat in well-separated feeding episodes, but during these episodes they exhibit short pauses in food-taking. A criterion was required to distinguish such intra-meal pauses, followed by the continuation of the same meal, from longer pauses indicating the termination of a meal and followed by the start of a new meal. A statistical analysis of the distribution of all pauses and interval durations in 200 daily recordings showed a regular gap between groups of short pauses of 30–40 min. So, a period of 30–40 min of no eating was taken as the best criterion for separating the basic prandial events (Le Magnen & Devos, 1980*a*; Devos, 1981). Thus, a meal is defined as a period of eating preceded, and followed by, at least 30 or 40 min of no eating.

This technique employs solid food and only a single type of food is offered at any one time. In other procedures, rats are presented with a choice between two or more diets, each in a separate food-cup, as in the classic investigations of self-selection of diets by C. P. Richter. More recently, in studies of rats offered a multiple choice of highly palatable foods termed 'cafeteria regimen', the use of food-cup-weighing recorders in each cage has permitted detailed study of the feeding patterns (Rogers & Blundell, 1980; J. Le Magnen & M. Devos, unpublished).

In many experiments, liquid diets are used instead of, or in addition to, solid food. The short-term measurement of drinking is achieved by recording the licking pattern through devices termed 'drinkometers' (Stellar & Hill, 1952). Such recording of licks allows the investigator to study the microstructure of the oral intake of a rat on a liquid diet (Allison & Castellan, 1970). It had never been possible previously to examine this microstructure of food intake in rats by recording chewing and swallowing pattern.

Self-intragastric and self-intravenous feeding

These procedures described above concern the oral intake of foods. Other techniques have been developed which allow us to study the self-intragastric or self-intravenous intake of nutriments in rats, monkeys and even in human subjects. In these sophisticated but now routinely used techniques, the self-injection pattern is assessed by recording the lever-pressing which instigates the delivery of a specified amount of the experimental solution (Epstein & Teitelbaum, 1962; Snowdon, 1969; Rowland *et al.*, 1975; Jouhaneau & Le Magnen, 1980).

Parallel measurements in unaltered conditions

The feeding response of animal models may be measured with simultaneous control of other variables. In addition to the strain, age, and sex of rats, the body weight is a major parameter. Body weights are taken daily, generally when the cages are cleaned (preferably at the start of the day). Rats are always housed in individual cages. Ambient temperature is kept either in a constant neutral zone (29 °C) or generally just below (21 °C). A factor to be controlled is the dark–light cycle which profoundly affects the daily feeding and underlying neurometabolic pattern. Unless otherwise prescribed by the procedure, water is available and its intake recorded by the reading of graduated drinking tubes or, as mentioned above, by automated drinkometer recording.

A decisive advance in studying the involvement of feeding mechanisms in body energy homeostasis has been obtained through a technique in which the two components of energy balance (food intake and energy expenditure) are measured simultaneously in rats over time. This technique, initially developed in the late 1950s, has been described in detail elsewhere (Le Magnen & Devos, 1970; Le Magnen *et al.*, 1973). Briefly, the rat is placed in a closed chamber in which dry air is pumped at a flow rate of 1.7 litre/min. Oxygen and CO_2 concentrations are measured in the outflow by gas analysers, printed out, and eventually recorded and processed in parallel by a computer. Within the chamber, the rat has free access to its food-cup, which is inserted in the strain-gauge weighing device. Hence its free-feeding pattern or short-term feeding responses are recorded in parallel with fluctuations in respiratory parameters. Through this indirect calorimetry and taking into account the intake of metabolizable energy, the balance between energy intake and energy expenditure may be evaluated. In addition alterations in the $CO_2:O_2$ ratio (respiratory quotient) with food intake yields insights into food metabolism.

Since the historical beginning of works on hunger, gastric secretions and contractions have been studied and assessed in relation to hunger sensations and/or food intake. The use of an air-inflated balloon to record gastric contractions led to a famous amusing methodological error. It was believed that the balloon recorded contractions of an empty and 'hungry' stomach. It was later recognized that the contractions were in fact elicited by the balloon in an attempt by the stomach to digest it! More recently, rates of gastric emptying and intestinal transit and absorption have been studied as possible variables controlling meal size (Newman & Booth, 1981; Kalogeris *et al.*, 1983; Reidelberger *et al.*, 1983).

Another advance has been provided by taking parallel recordings of food intake and of the blood concentration of various metabolites and glucoregulatory hormones in free-moving rats. This technique uses chronically implanted cardiac catheters (Steffens, 1969*a*) and home cages equipped for such rats (Nicolaïdis *et al.*, 1974). Initially, rat blood samples were withdrawn every 10 min and free intake was recorded for 1 or 2 h (Steffens, 1969*b*). A more refined technique has been developed, in which a microflow of blood (0.9 ml/h) is withdrawn continuously throughout 12 h or more (Louis-Sylvestre & Le Magnen, 1980*a*). Replacement blood from a donor is given continuously to the experimental rats through a femoral catheter. The food intake *ad libitum* is automatically recorded by the food-cup-weighing device. The blood glucose level is continuously measured in the blood outflow by a glucose autoanalyser. In other experiments, using the same technique, insulin, glucagon or catecholamine levels can be determined on successive 2 min blocks of blood samples. Intrachronic intraportal and/or sus-hepatic venous catheters have also been used (Strubbe & Steffens, 1977*a*; Langhans *et al.*, 1982).

Food intake and parallel measurements in normal human subjects

All the above techniques are applied to animal models, mainly the rat. But what about human subjects?

The on-going control of daily voluntary food intake under laboratory or hospital conditions is a difficult problem. Various types of inquiries and interviews, particularly with obese people, have been proposed and discussed (Mayer *et al.*, 1965; Huenemann, 1972; Rolland-Cachera *et al.*, 1983). Restaurants and particularly self-service cafeterias are suitable for discrete observation of the choice of foods and amounts eaten (Coll *et al.*, 1979). In a canteen, it is even possible for an observer to record the microstructure (chewing and swallowing) of a consumer with a multichannel event recorder (Warner & Balagura, 1975).

Under laboratory conditions, the recording of food-taking from the plate, and in successive courses, has been carried out, as with rats, by a continuous monitoring of the weight of the plate by a computer (Kissileff *et al.*, 1980). Parameters such as the eating rate from the beginning to the end of the meal can be measured with this 'feeding machine'. A further step in the microanalysis of man's eating behaviour has been reached by the development of a technique for recording the chewing–swallowing pattern (Pierson & Le Magnen, 1970; Bellisle & Le Magnen, 1980). Subjects are offered a meal composed of bits of bread of constant volume and size. A layer of various items (butter, pâté, jam, etc....) makes these pieces of bread palatable in different ways. A graphic recording of masticatory and swallowing movements, by sensitive equipment, allows the experimenter to analyse the human meal by using a series of parameters of the chewing–swallowing pattern.

Fortunately (or unfortunately), man can speak. Psychophysical methods have been applied to evaluate the intensity of hunger and satiety, or of pleasantness or unpleasantness of tasted foods. Category scaling, magnitude estimations or visual analogue ratings have been used as procedures in hunger and satiety ratings before and after a meal. Similar procedures have been extensively used to rate the level of pleasantness–unpleasantness as a function of the concentration of a particular tastant or odorant in a solution.

Such psychophysical studies must be used with caution. Generally, the evaluation of 'pleasantness' is performed by tasting a sample of the item but not swallowing it. Such a judgement of pleasantness is a misleading test of *palatability*, i.e. the ingestive response to the orosensory activity of a food. A 20% sucrose solution judged pleasant by tasting is judged strongly distasteful by the subject asked to drink 50 ml of this solution (F. Bellisle, unpublished).

Self-intragastric feeding has been investigated and measured in man (Jordan *et al.*, 1966). Subjects press a button and in so doing freely command the direct delivery of a liquid food into the stomach via an oesophageal tube. The flow rate is determined by the experimenter. The amount self-injected in a given state of the subject is measured, and compared with the amount drunk by controlled delivery through the mouth. Simultaneous oral and gastric deliveries at different and varying flow rates can also be studied using this technique.

Direct or indirect calorimetry has been used for people who have normal weight or are obese, particularly in the study of body energy homeostasis (Jequier, 1982/1983). The daily ratio of energy intake to energy output, and the heat increment of food, can be measured in subjects staying 24 h or more in a calorimetric chamber. In normal healthy subjects, blood parameters can also be measured in relation to hunger and satiety and to the intake of food. Twenty years ago, differences in arteriovenous blood glucose concentrations before and after a meal were measured by blood sampling (Mayer, 1953; Stunkard & Wolff, 1954). Now, under medical supervision, volunteers have changes in their blood glucose, insulin, and glucagon levels recorded for 1–2 h during food-taking (Bellisle *et al.*, 1983).

An original technique has been developed to test preferences and aversions in human neonates (Steiner, 1973), and has also been applied to the rat (Grill & Norgren, 1978*a*). It consists if a video-tape recording of the facial expressions of babies stimulated by olfactory or gustatory food-related stimuli. Rejection and acceptance are clearly distinguished.

Experimental manipulations

The above techniques and procedures developed to assess the intake of food may be used *per se* in normal subjects, whether or not other measurements are taken at the same time. The investigation of the mechanisms which govern food intake requires that there is an association between the same basic measurements of intake and various experimental manipulations of the normal state. These manipulations may affect the food, the gastrointestinal tract, the liver, the systemic compartment and blood contents, the adipose tissues (white and brown), and finally and essentially the brain and the somesthetic as well as the autonomic nervous system.

Diets

The short- or long-term free intake of many types of experimental diet may be compared with the intake of the complete stock-diet. Experimental diets differ from this control diet by changes in nutritive properties or in some sensory activities (smell, taste, texture) or both. The response to high fat diets (40–50% fats) has been studied. In addition to high fat content, this diet differs from the stock-diet in two ways: (*a*) its caloric density has 66% of calories provided by fats and (*b*) it has a low carbohydrate content. The respective effects of these simultaneous changes are not often clearly dissociated. In order to eliminate the effect of the high caloric density, a low caloric high fat diet has been prepared by addition of inert material (cellulose). High, low and free protein diets, and diets lacking only one essential amino-acid, have been prepared so that specific appetite for protein may be studied. Likewise vitamin-, Ca^{2+}- and Zn^{2+}-free diets, for example, have been used to test specific appetites for vitamins or minerals exhibited by animals deficient in these respects. Without changing the proportions of the three major macronutrients, the caloric density of the stock-diet is easily manipulated by the addition of non-metabolizable materials – cellulose or kaolin. Liquid diets of varying caloric densities have also been used to study the adjustment of caloric intake in man (Campbell *et al.*, 1971; Spiegel, 1973).

Various techniques and procedures have been used to dissociate the respective roles of sensory and nutritive properties of diets in the control of intake. The simplest procedure is the comparison of responses (1) to two diets identical in their sensory activities and differing in their nutritive properties or (2) to two diets identical in their nutritive properties and differing in their sensory activities.

The comparison of intake of saccharin solution to that of sucrose or glucose solution is an example of the first type of response which has been highly fruitful in many investigations (Le Magnen, 1977*a*). A comparison of intake of solutions of NaCl and of LiCl has been also made (Nachman, 1963). Comparison of the course of intake of a quinine-adulterated toxic diet with that of a sucrose octoacetate-adulterated non-toxic diet is based upon the same rationale (Kratz *et al.*, 1978; Aravich & Sclafani, 1980).

Odour labelling of diets was first used in the 1930s (Harris *et al.*, 1933) to demonstrate learning of specific appetites on the basis of olfactory cues. The same procedure was used later by various researchers to demonstrate the learning of palatability induced by the post-ingestive nutritional activity of a food (Le Magnen, 1956; Booth, 1972*a*).

Another means of evaluating the role of sensory analysis of foods is to eliminate the sensory organ functions surgically. A complete ageusia is difficult to introduce by sectioning of the chorda tympani, and of the glosso-pharyngeal and pharyngeal branches of the vagus without impairing simultaneously the trigeminal and other non-gustatory afferents. Anosmia in rats is achieved either by a topical application of $ZnSO_4$ on olfactory mucosa or by surgical olfactory bulb ablations. The first method may give misleading results. It has been shown that regeneration occurs rapidly: olfactory responses are recovered after only 4–5 days (Larue, 1973). This reversibility of anosmia can be exploited.

Gastrointestinal negative feedback

Many techniques, surgical and otherwise, are used to test the role of the alimentary canal in determining the short-term food intake. The dissociation of (*a*) facilitatory and inhibitory actions of the food passing into the mouth from (*b*) the counteracting action of the stomach and intestine is obtained in various ways.

In rats, a condition of sham-feeding, i.e. of oral intake of foods which go out through an open fistula and then do not enter the stomach, is achieved by oesophagotomy (Mook, 1963) or by a gastric open cannula (Liebling *et al.*, 1975; Deutsch *et al.*, 1980). In the latter method, closing the cannula allows the experimenter to re-establish the normal oro-gastrointestinal transit for comparison (Antin *et al.*, 1977; Kraly *et al.*, 1978). Sham-feeding by mouth can be associated with infusions of a liquid food to the stomach and intestine or of various solutions which may or may not differ from what is being orally sham-fed. They are pumped at various flow rates, either before, during or after the allowed sham-feeding. In some experiments, foods entering the stomach can be withdrawn through a gastric fistula or, conversely, a solution can be added to the stomach content (Snowdon, 1969; Deutsch & Gonzalez, 1980). In others, a ligated pylorus prevents gastric emptying to the intestine (Kraly *et al.*, 1978). A detailed description of these techniques, particularly of the surgical operations and of cannula placements, will be found in the references cited here.

These methods, generally used to investigate the role of combined oral and gastrointestinal actions in determining the amount eaten during one meal, are also used to study meal patterns in chronic conditions. The distinction between factors acting respectively on meal size and post-meal satiety duration is then important and unfortunately often ignored.

In chronic experiments of tube-feeding, two types of chronically implanted catheters have been used: the nasopharyngeal catheter (Epstein & Teitelbaum, 1962) and the chronic gastric cannula. The last one allows tube-feeding to be performed without interruption for one month or more (Nicolaïdis *et al.*, 1974).

Systemic alterations

Acute and chronic changes of the *milieu intérieur* and their effects on food intake can be investigated by injecting or infusing various agents or drugs. The dramatic technique of the chronic intracardiac catheter, previously mentioned with respect to its use for blood withdrawal, allows the investigators to perform long-term infusions in free-moving rats (Rowland *et al.*, 1975; Larue-Achagiotis & Le Magnen, 1979).

Various organ ablations are used to alter the internal and metabolic state of animals and to study the effects of this alteration on food intake: hypophysectomy, adrenalectomy, partial or complete pancreatectomy, etc. Feeding in diabetic rats can be studied by first inducing diabetes with streptozotocin. In such diabetic rats, kidney or liver transplantation of pancreatic islets has been used to test the role of selective denervation of beta-cells (Louis-Sylvestre, 1978*a*; Inoue *et al.*, 1978). Porto-caval shunts by-passing the liver have been used to study and to eliminate a role of the liver in the meal patterning (Louis-Sylvestre *et al.*, 1979). Finally, parabiosis between pairs of rats and their food intake being separately controlled, has been used to good effect (Schmidt & Andik, 1969; Hervey *et al.*, 1977; Parameswaran *et al.*, 1977). To improve upon this parabiosis by graft, cross-circulation by an external peristaltic pump, between two partners has been attempted. Although successful in the monkey (Walike & Smith, 1972), it has not as yet been achieved in a smaller animal such as the rat.

Manipulation of adipose tissue lipogenesis and lipolysis can be achieved by lipogenic and lipolytic drugs or other agents. In addition, lipectomy (reaching 45% of the subcutaneous adipocytes in rats) has been used (Faust *et al.*, 1977). In rats and other animal models, excision of brown adipose tissue involved in regulatory thermogenesis has been tried (Stephens, 1981).

In man, most of these systemic manipulations must be excluded for obvious ethical reasons. However, taking advantage of surgical operations necessary for other purposes, some experiments have been made, for example, in adrenalectomized patients or in states of diseases such as diabetes mellitus, liver diseases, etc.

The most evident correlates of feeding, i.e. energy imbalances, can be examined by various procedures. In addition to food deprivation, or, conversely, forced overeating (performed by tube-feeding), body energy imbalance is achieved by altering thermogenesis. Rats are forced to exercise by swimming or on a tread-mill or run-wheel (Mayer *et al.*, 1954; Thomas & Miller, 1958; Stevenson *et al.*, 1966). Thermogenesis is increased by cold exposure, which elicits an increase of food intake in which the respective roles of cooling the skin and of energy losses can be investigated (Brobeck, 1948).

Central and peripheral nervous system

A considerable number of experimental studies on food intake (not surprisingly) involve actions on the peripheral nervous system and on the brain. In the former, various nerve sections, neural discharge recordings or electrical stimulations may be investigated. The vagus and splanchnic nerves are most frequently involved (Kral *et al.*, 1983). In rat, surgical techniques allow experimenters to perform a full subdiaphragmatic vagotomy and a selective gastric vagotomy. Various post-operative tests have been proposed (and opposed) to verify the completeness of vagotomy (see review in Louis-Sylvestre, 1983*a*). Sympathetic and/or parasympathetic denervations of the liver in rats and dogs can be performed easily. Here also, tests of completeness of the denervation are used, such as the elimination of heart responses after intra-portal injections of glucose (Louis-Sylvestre *et al.*, 1980).

Discharges in afferent fibres to pancreas, liver and adrenals can be recorded in anaesthetized animals and changes in their states of nutritional imbalance can be monitored (Niijima, 1975). Discharges of vagal afferent fibres to the brain can also be recorded, and have yielded evidence of the presence of chemo- and mechano-sensors in the stomach and intestine (Mei, 1978; Jeanningros & Mei, 1980).

At brain level, various classical techniques have been employed to explore the cerebral mechanisms which govern feeding. Only some specific techniques will be mentioned here.

In a small number of studies, parallel recordings of free intake of food and of brain electrical discharges have been performed by employing electrodes chronically implanted in critical brain sites such as the lateral hypothalamus (LH) (Hamburg, 1971; Ono *et al.*, 1980). Single-unit recordings have been successfully used in wide-awake monkeys (Rolls & Rolls, 1981). Multi-unit and single-unit electrophysiological recordings in the olfactory bulb have been also performed in rats and rabbits before, during, and following a free oral intake of a meal (Pager *et al.*, 1972). Identical recordings in peripheral gustatory nerves or in the nucleus tractus solitarius have not been achieved except under stimulation of the tongue by solutions which can be tasted (Glenn & Erickson, 1976). Electrolytic lesions or knife-cuts in the ventromedial hypothalamus (VMH) and LH have been the subject of several thousand publications. Chemical neurotoxins such as kainic acid (Grossman *et al.*, 1978; Stricker *et al.*, 1978), which causes lesions limited to cell bodies, are injected via implanted cannulas (Myers *et al.*, 1967; Walls & Wishart, 1977). These cannulas are also used for acute or chronic intraventricular or local injections. In particular the neurotoxin gold-thioglucose (GTG) can be used to obtain necrosis of the glucose-sensitive neurones of the VMH (Mayer & Marshall, 1956). Hyperphagia and obesity are induced in mice by 1 ml/g of body weight of this drug systemically administered. Glucose antimetabolites such as 2-deoxy-glucose also have specific uses: injected systemically or intraventricularly, they stimulate feeding. Finally, brain electrical stimulations and self-stimulations are used. The former techniques have been adapted to unanaesthetized and free-moving animals.

2 Basic facts and normal feeding

Feeding behaviour, in its broader sense, comprises two different components, which are the subjects of separate but nevertheless complementary fields of investigation. In the wild, animal selection and ingestion of foods that are required for growth and maintenance are preceded by and dependent on food-seeking behaviour. This ecological or ethological aspect of feeding behaviour belongs to Natural History. To make a descriptive or experimental study of feeding strategies needs control and isolation (which is difficult to achieve) of a tremendous number of variables in the environment. These variables can be controlled by the experimenter but cannot be modified or artificially kept constant. The physiological mechanism which governs the feeding behaviour proper and pushes animals to seek and to eat food is the basic determinant of this food seeking. In the laboratory, ecological variables are necessarily excluded by the experimenter in his study of feeding responses in 'experimental situations'. In such circumstances, only variables which directly affect the physiological process can be controlled, kept constant or manipulated, one at a time. The experimental analysis of feeding in captive animals yields the basic knowledge which enables us to return to an ecological approach.

Basic concepts

Feeding, associated with drinking and respiratory intake, is the behavioural component of a physiological process – nutrition. Food selection and amounts eaten are responses to two types of metabolic demand. Supplies of carbohydrates, fats and proteins as interchangeable sources of metabolizable energy are required to cover energy expenditure. Simultaneously, essential amino-acids, vitamins and minerals must be provided by feeding in proportion to their rates of catabolism.

Current energy losses of the body, which are measured by direct or indirect calorimetry, include the basal metabolic rate and three categories of additional expenditure: muscular exercise, regulatory thermogenesis, and the heat increment of foods. The total energy metabolism fluctuates widely over time according to the amount of heat produced in excess of the basal metabolism. At cellular and whole-body levels, this energy output is continuous. It requires an uninterrupted supply to the tissues of oxygen and energy metabolites (or fuel). Both supplies are dependent on intake from the environment and therefore on behaviour patterns. However, the intermediary steps between intake from the environment and supply to the tissues are very different in respiration and in feeding. The consequent dependency of the tissue supply on intake is likewise different.

Blood is the dispenser of oxygen and fuel to the tissues. Both outputs from the blood are compensated for by endogenous mechanisms which control and regulate input into the blood. This regulated input/output maintains a constant P_{O_2} on the one hand, and on the other a constant level of one of the energy-yielding metabolites, namely glucose. But the means of loading oxygen and fuel into the blood and the mechanisms of its regulation are very different. In the absence of substantial oxygen stores in terrestrial animals, the blood oxygen content is maintained from moment to moment and is dependent on continuous respiratory intake. Instantaneous regulation of fluctuations in tissue oxygen consumption is also achieved by pulmonary ventilation through the sensor of the carotid sinus. Because foods are not as readily available in the environment as is air, a similar system could not achieve the continuous supply of fuel to tissues and the maintenance of blood glucose concentration. Feeding, in all animal species, is essentially periodic in nature. This discontinuous intake, in contrast to continuous expenditure, is made possible because the loading is drawn from body energy stores. The continuous feeding of the blood dispenser is regulated by hormonal and neural control of both hepatic outflow and fat mobilization.

Three types of store of either large or small capacity are involved: a small hepatic and muscular glycogen store, a gastrointestinal store loaded by each oral intake, and the large capacity store of body fats. The store of body fats, 12–15% of the total body mass, allows rats and humans to survive weeks or months of starvation.

From experimental studies which will be detailed in Chapter 3, it has been concluded that the intake of food not used directly in the supply of fuel to tissues and in its regulation is involved in filling these stores periodically in response to their depletion.

Normal feeding in the steady-state condition

The ultimate goals of experimental investigations on feeding are first to separate and then to identify mechanisms by which an animal:

(1) is stimulated to eat;
(2) eats a given amount of a particular food during a bout of eating or meal, until satiation;

(3) is not stimulated to eat or is satiated thereafter;
(4) exhibits a middle- and a long-term cumulative intake which balances its various metabolic demands.

The first step of the investigation is controlling feeding by recording the eating patterns of an animal model such as the rat, in a series of standardized conditions.

Free-feeding pattern in rats

The free- or *ad libitum* feeding pattern of rats is the basic condition. Water being available at all times, rats may be offered free access to a familiar complete mixed diet in excess of normal requirements. Their *ad libitum* intake is recorded by means of the techniques described in Chapter 1. This standardized condition eliminates some variables or controls others. Energy expenditure resulting from activity or affected by ambient temperature is kept constant. Energy retention due to growth is eliminated by using fully-grown adult animals. The constant free access eliminates the acute effect of deprivation on feeding responses (see below). The food is familiar. This eliminates both neophobia and the difference in palatability between previously experienced and unexperienced foods. Finally, this diet is mixed and nutritionally balanced, thus excluding the free selection of macronutrients and therefore the manifestation of nutrient-specific appetites.

What is observed in this condition? There is a dual periodicity (Fig.

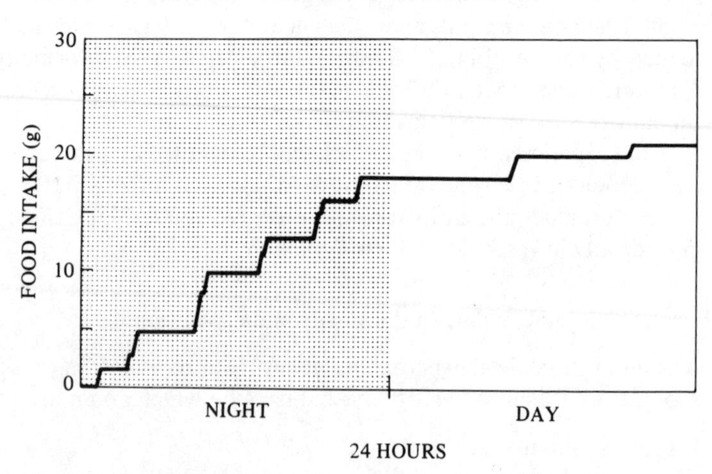

Fig. 2.1. Recorded daily feeding pattern of rats fed *ad libitum*. Prandial and circadian periodicities may be seen clearly.

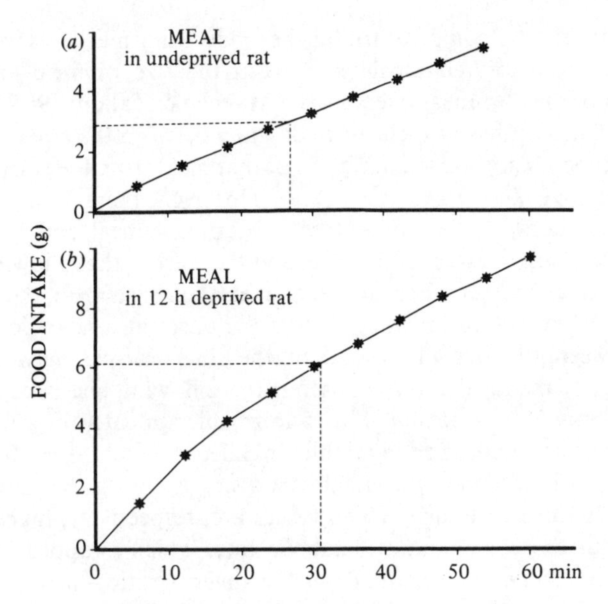

Fig. 2.2. Intake rates during (*a*) a long, large nocturnal meal in undeprived rats and (*b*) during the first meal after 12 h of food deprivation. A constant rate is observed from the beginning to the end of meal in the *ad libitum* condition. With food deprivation, the eating rate is accelerated during the first half of the duration of the meal.

2.1). The 'prandial' periodicity is clear. Rats take discrete meals. These meals can be easily identified by the criterion of a minimal duration of pre- and post-meal intervals (see Chapter 1). From the initiation to the end of the meal, the amount eaten, or meal size, varies from 1 to 6 g. Except for a slight acceleration at the beginning of the meal rats eat at a constant rate of 0.2 g/min during both night and day (Fig. 2.2). The meals are separated by intervals of no eating which differ largely in duration from 40 min to 4–6 h or more. A circadian or nycthemeral periodicity is superimposed on this basic prandial periodicity. Rats eat bigger and longer meals during the dark than during the light period. Larger meals at night are separated by shorter intervals than are the smaller daytime meals, so that the nocturnal cumulative intake amounts to 75–95% of the 24 h intake. No periodicity greater than 24 h is apparent in male rats. On 200 daily recordings (20 consecutive days in 10 rats), no positive or negative correlation was found between intakes on successive days (Le Magnen & Devos, 1980*a*; Devos, 1981). In females, the 4–5 day oestrous cycle gives rise to an equivalent cycle of feeding, with a fall of intake on the day of oestrus (Wang, 1923; Slonaker, 1925).

An important finding regarding the underlying mechanisms came from the study of relationships between the size of meal and the duration of meal-to-meal intervals (Le Magnen & Tallon, 1963, 1966). No correlation exists at night or in daytime between the size of meals and the time separating the initiation of that meal from the start of the preceding one (*pre-prandial interval*). However, there is a positive correlation between the amount eaten during a meal and the time elapsed between the start of the meal and the start of the following one (*post-prandial interval*). The more the rat eats during a meal, the longer the state of no eating or satiety before the onset of a new meal. The link between the size of meals and the time elapsed between the beginning of the meal and the onset of the following one is known as the *post-prandial correlation*. The ratio of calories eaten in a meal to the post-prandial duration in minutes is called the *onset ratio*. In adult rats, this ratio is about 70 cal/min on average during the night and 20 cal/min during the day. These values are, respectively, higher and lower than the concomitant metabolic rate, which is approximately 45 cal/min. Some investigators use the inverse ratio – post-prandial duration to preceding meal size in grammes, designated by them the *satiety ratio* (Booth, 1972*a*; Panksepp, 1973). This post-prandial correlation, initially found in rats and confirmed by many investigators in this species (De Castro, 1975; Bernstein, 1976*a*; Davies, 1977), has also been found in various other species: monkey (Natelson & Bonbright, 1978; Hansen *et al.*, 1981), dog (Ardisson *et al.*, 1981), rabbit (Sanderson & Vanderweele, 1975), chick, goose, and humming-bird (Duncan *et al.*, 1970; Wolf & Hainsworth, 1977; Marcilloux, 1980). It is absent in guinea-pig (Hirsch, 1973), and doubtful in cat and pig (Mugford, 1977; Auffray & Marcilloux, 1983).

In rats, this correlation is high during the night, reaching $r = 0.90$ in some rats, and is highly significant. The inter-individual/intra-individual correlation established over 200 days in a particular group of 10 rats is illustrated in Fig. 2.3. For the light period, small meals are followed by long intervals before the start of the next meal. This daytime post-meal interval, in contrast to that of the night-time, is poorly or not correlated to the size of the meal (Fig. 2.3). The regression and the equation of the linear relationship indicate that, after a meal of the same size, the rat remains satiated on average 2.5 times as long during the day as during the night.

What is the meaning of relationships regarding the underlying mechanisms which determine meal initiation and meal size? The relationships between meal size and meal-to-meal intervals suggest two separate and nevertheless interacting mechanisms governing meal initiation and its frequency on the one hand, and meal size on the other.

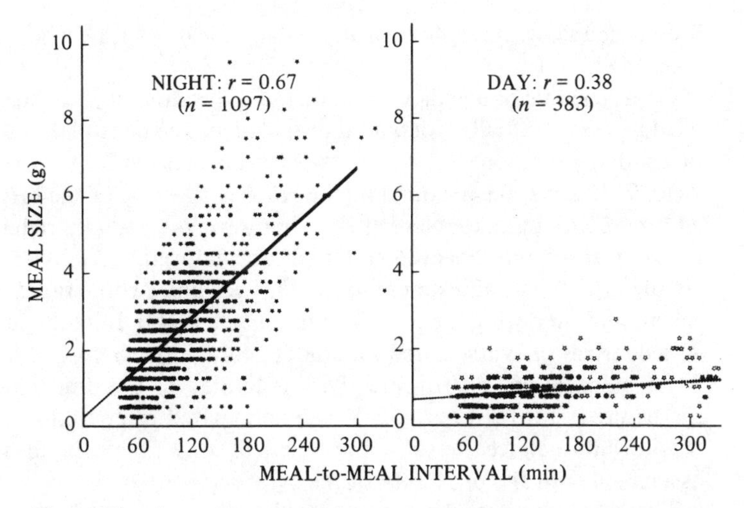

Fig. 2.3. There is a positive correlation between meal sizes and post-meal-to-meal intervals during the night and the daytime in free-fed rats. n = number of meals.

A meal is initiated after a time which is proportional to the caloric load achieved in the previous meal. This suggests that the triggering mechanism for the start of a meal or stimulation to eat involves an all-or-none signal resulting from the rate of metabolic utilization of the amount of food ingested in the preceding meal. Consequently, the difference in meal size/post-meal interval ratio between night and daytime, the origin of the circadian periodicity, may be due to a difference in the rate of food metabolism during the two periods. A lower rate during the daytime should explain the delay of meal onsets relative to the size of the preceding meal.

Thus, the time of meal initiation and, therefore, also meal frequency are dependent on preceding events. The amount eaten from this initiation to satiation (or meal size) is not similarly dependent since it cannot be correlated either to the pre-meal interval or to the preceding meal size. It is therefore suggested that in this *ad libitum* condition the mechanism which determines the size of meal is not affected by, nor is a function of, the state which has initiated that meal. Rather, the amount eaten within a meal becomes a store in anticipation of subsequent expenditures. This suggestion will be fully confirmed later by identification of the mechanisms involved.

Feeding schedules

In a scheduled feeding, free access to food is eliminated. Foods are offered once daily for 1, 2, 4 or 6 h and this one daily meal is regularly offered at the same time either during the night or during the day. A fixed schedule of three, four or more daily meals of 1 h each, has also been tried.

On a schedule of one daily meal of 1–4 h duration, intake during the meal increases initially, compared to the *ad libitum* control, as an effect of food deprivation. Then meal sizes keep progressively increasing. After 7–10 days, the maximal intake reaches 50–70% of the preceding 24 h *ad libitum* intake. Nevertheless body weight, initially reduced, is re-established, and rats even gain their normal weight 2 to 3 g per day (Baillie, 1977). The adjustment to 2 h daily feeding is more rapid during the night than during the day (Balagura *et al.*, 1975). Initially the 20 to 23 h deprivation is associated with the concomitant increases in activity. After habituation, this hyperactivity is limited to the time preceding the habitual meal (Evans, 1971). The anticipated response to the time of the habitual feeding is conditioned to this time and to stimuli associated with the proximity of the meal.

These data support the notion that the increase of meal size in such feeding schedules is a learned anticipatory response. As an effect of this learning, rats coming back to feeding *ad libitum* overeat transiently (Lawrence & Mason, 1955). This anticipatory or provisional appetite is supported by evidence from the following experiment (Le Magnen, 1959*b*): rats are habituated to a feeding schedule of three 1 h meals per day, presented every 7 h. After habituation, one of the three meals was suddenly omitted. Initially the size of meal occurring after the new 15 h deprivation increased; but progressively the size of meal preceding the 15 h gap also increased. After 12 days, 75% of the amount previously eaten in the omitted meal were added to the pre-fast meal.

Normal feeding in man

In man, in a steady-state condition with respect to mean energy expenditure and food availability, the daily food intake is realized in the habitual schedule of three to four diurnal meals at fixed hours. In the absence of a possible regulation by changing meal-to-meal intervals, only a short- or long-term adjustment of meal sizes is effective to match energy intake and expenditure. A negative correlation between the amount eaten at lunch and at dinner has been reported (Jiang & Hunt, 1983). When, as a result of isolation, man lacks temporal cues, his feeding pattern becomes similar to the free-feeding pattern of rats. Human subjects cut off from any contact with the outside for some days, and thus totally deprived of temporal cues, ate progressively in a 'free-running' condition. Then, like rats in *ad libitum* feeding, the post-prandial correlation was observed. These voluntary prisoners asked for their next meal to be served after a time proportional to the amount eaten in the previous meal (Bernstein, 1981).

Does the sensation of hunger and its degree play a role in the normal condition of meals taken daily at the habitual hour? As in the rat placed in an equivalent feeding schedule, external stimuli act as conditioned stimuli of systemic changes which, in turn, produce or exaggerate the state of the central nervous system (CNS) associated with feeling hunger (see Chapter 3). However, there is no indication that the intensity and even the presence of a perception of hunger are involved in determining either the size or the time of initiation of the habitual meal. Hunger level at the time of a meal can be assessed by hunger ratings. Clearly, this score does not predict the amount eaten during the complex human meal, which is mainly dependent on eating habits and varies in the three daily meals – breakfast, lunch and dinner – within the sociocultural context. When different courses are presented, sensory-specific appetite and sensory-specific satiety are the major determinants of amounts eaten (see Chapters 3 and 4).

The microstructure of the human meal has been studied under standardized and experimental conditions. Recording of the chewing–swallowing pattern, as in studies using a feeding machine (see Chapter 1), shows a typical evolution of eating rate from the beginning to the end of a meal. A mathematical analysis of these curves shows that the best fit is obtained with a quadratic equation (Bellisle & Le Magnen, 1980; Kissileff *et al.*, 1982).

Normal feeding in the non-steady-state condition

Food deprivation

The state of energy deficit and weight loss induced by food deprivation is the most obvious condition by which hunger and eating are stimulated. Before examining the mechanisms of this stimulation, it is useful to study the feeding response to deprivation and its relation to the level of energy deficit assessed by weight loss or by other means.

Symptoms of hunger under food deprivation have been sought in animal models. Increase in activity as a function of duration of fasting, increasing rate of lever-pressing for food in a variable interval schedule, and strength of obstruction overcome by rats to obtain foods, have all been used and presented as these 'symptoms of hunger' (Varner, 1928; Heron & Skinner, 1937). There is some agreement between results from these various experiments for a progressive increase in this 'urge to eat' for up to 5 days of total fasting in rats, followed by a fall in the symptoms. However, there is notable general disagreement between the evaluations of the intensity of hunger arousal and food intake under food deprivation. It is not surprising since (as will be discussed in Chapter 3) the initial stimulation to eat is only one among many factors

– e.g. palatability of the food, process of satiation – which affect amounts eaten.

Rather than consider the amounts eaten at the restoration of food access, latency or readiness to eat has been measured as a function of deprivation from 0 to 92 h in rats. This latency decreases as a power function of the duration of deprivation. The percentage of weight loss compared to the weight reached by fed controls and the duration of fasting were similarly related. Thus, latency decreases directly with the percentage weight loss (Bolles, 1962).

When rats are deprived of food for less than 24 h, the feeding and metabolic circadian cycle strongly interferes with the effect of deprivation on re-feeding. The times of day when rats are deprived and when their response to the fast is tested are important variables. They are difficult to dissociate when, for example, deprivation is achieved during the night and the response is tested during daytime or vice versa (Bare & Cicalla, 1960; Bellinger & Mendel, 1975). A comparison of responses to short-term food removal both achieved and tested either at night or in the daytime yields evidence of a decisive role of the metabolic background in this deprivation-induced feeding (Le Magnen *et al.*, 1980*a*) (Fig. 2.4). A food withdrawal from the beginning of the night for 2, 4, 6, . . . h, leads to a linear increase in the size of the first meal at the restoration of food access. Thus, the size of meal, unaffected by the duration of the pre-meal period of no eating in the *ad libitum* condition becomes highly dependent on this pre-prandial interval of no eating when it is prolonged by the removal of food access. At the beginning of the night, a prolongation by about 50% of the normal between-meal interval by food removal between the first and the second meals is sufficient to induce a significant increase in the delayed second meal (J. Le Magnen & M. Devos, unpublished). If, following the big initial meal in the re-feeding period, the occurrence of subsequent meals is accelerated, a correlation between meal sizes and post-meal interval duration is still observed, but the meal size : meal-to-meal interval ratios are considerably elevated (Le Magnen, 1977*b*). The effect of the same schedule of deprivation during daytime is profoundly different. The same increment in the first meal after deprivation as observed at night is obtained by prolonging the pre-meal no eating period for twice as long as was observed at night. For example, an increase of intake equivalent to that induced by a 3 h-deprivation at night is obtained by a more than 6 h-deprivation during the daytime (Le Magnen *et al.*, 1980*a*). Deprivation for 6 h at the beginning of the night doubles the intake in the subsequent 6 h. Deprivation for 6 h at the beginning of the day (the time at which rats eat very little in the *ad libitum* condition) is without effect on the last 6 h intake in daytime.

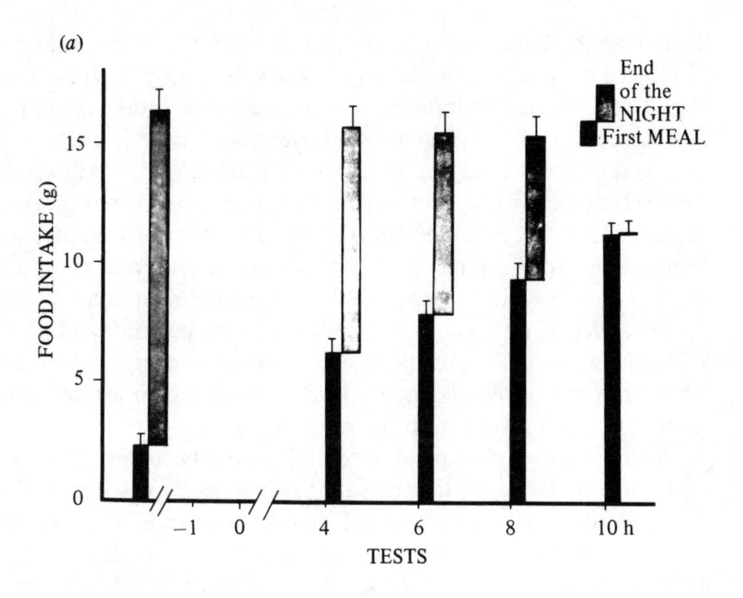

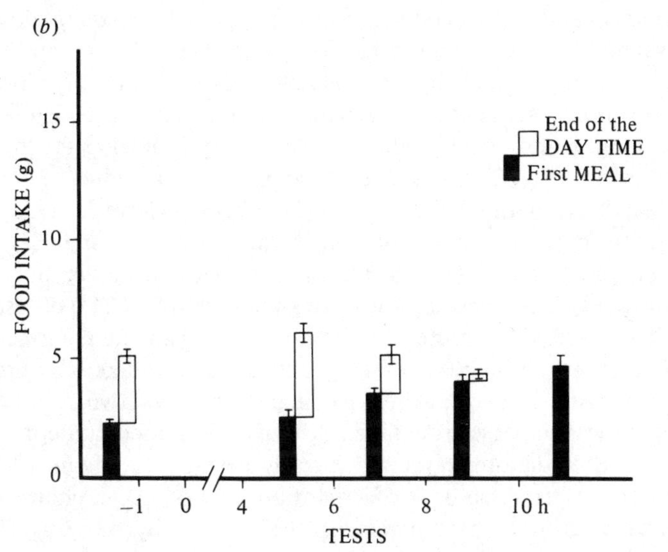

Fig. 2.4. Increases of the first meal size and subsequent intake, until the end of the period, after the restitution of food following progressively increasing intervals of food deprivation.

Responses to 12 h nocturnal and daytime fasts are different. Following 12 h of nocturnal fast, intake during the subsequent daytime is two or three times the normal intake. This high daytime intake is mainly due to a huge meal of 6–7 g eaten over more than 2 h at the restoration of food access. Conversely, a period of 12 h of food deprivation during the daytime has little effect on the subsequent nocturnal intake. One might expect to see a similar difference in man. Evidently, a food deprivation from 9 p.m. to 9 a.m. would be without effect on the daytime intake whereas deprivation of the daytime intake would have a strong effect on intake of food restored at the beginning of the night. We will return to the significance of these differences with respect to differences in the blood glucose level induced by either nocturnal or daytime food deprivation in the next chapter.

When food deprivation exceeds 24 h and becomes 'fasting', the rehabilitation by re-feeding the animal has been questioned. Contradictory results have been reported. In one study, rats were deprived of food for 1–4 days. Whatever the duration of fasting and the resulting weight loss, the increase of intake at the restoration of food access was identical and was equal to the preceding 24 h *ad libitum* intake. If food availability was restricted to the previous *ad libitum* intake during re-feeding, rats nevertheless re-established their weight (Levitsky *et al.*, 1976). The last result is in keeping with other reports which indicate a prolonged and strong reduction of activity following fasting in rats (Finger, 1951). The possible participation of a reduction in energy expenditure in the regulation of food intake could have been exaggerated by preventing the increase of intake in the above experiment. In another experiment, 24 h fasting in rats was reported to induce only a 20% increase in food consumption during the subsequent 24 h of restoration of food access. This increase was entirely due to an initial big meal. After 48 h fasting, an increase of 40% over the 24 h intake of controls was achieved during two days of re-feeding. During these two days, nocturnal intake was increased, daytime intake was decreased and consequently the night to daytime ratio was strongly elevated (Le Magnen & Tallon, 1968*b*). After some days of starvation, rats offered a choice between three macronutrients augmented their intake by selecting fats only (Andik *et al.*, 1951).

In as much as long-term fasting causes weight loss and the depletion of body fat mass, the correction of fasting during re-feeding involves liporegulatory mechanisms. This will be examined in Chapter 6.

The fact that food deprivation generates hunger sensation in man is self-evident. It is the most trivial and often the most tragic experience of humanity. In contrast to laboratory animals in which, as mentioned above, hunger arousal seems to increase until 4–5 days of fasting, the perceived intensity of hunger in man has been reported to reach a plateau after 2 days of total fasting and then to decrease (Langfeld, 1914). However, other reports indicate that a severe restriction, contrary to the total fast situation, maintains a continuous highly painful sensation of hunger (Berger & Le Magnen, 1960).

Historically, the belief that hunger was caused by gastric contractions or by other sensory afferent stimuli coming from the empty stomach comes from Aristotle* and not from Cannon & Washburn (1912), as it is claimed in many textbooks. Early on, it was shown that vagotomy does not eliminate hunger and eating after food deprivation (Janowitz & Ivy, 1949). It has now been fully demonstrated that the gastric component of the sensation is one of the *effects*, and not a cause, of the state induced in the central nervous system by the metabolic need for food.

Ambient temperature and exercise

Cold and heat exposure

Stimulation of feeding in conditions of provoked energy expenditures is one of the fundamental states used to further the understanding of regulatory mechanisms. In a range of ambient temperatures, designated the neutral zone, homeotherms can disperse their obligatory heat production due to their total metabolism. The temperature gradient permits a rate of heat dispersion compatible with the maintenance of body temperature. Below this range in cold ambient temperatures, the rate of heat loss becomes higher than the rate of the obligatory heat production. Three solutions are then possible and are realized in various species. In order to maintain the body temperature extra heat production occurs, and in order to spare the body fat store as fuel for this extra heat production, some animals increase their food intake proportionally and thus also maintain their body weight. Another possibility, achieved in other species, is the maintenance of body temperature by extra heat production in the cold but without, or with an insufficient increase in energy intake. This leads to weight loss which compensates a previous weight gain in some species during the cold season in a circa-annual seasonal weight gain/weight loss cycle. In a third case, the hibernators, the extra heat production is spared by the surrender of homeothermia. Nevertheless the hibernating animal balances its reduced energy expenditure by fat utilization associated with a partial or total anorexia. In the

* *Ethique à Nicomaque*, Librairie Philosophique VRIN, Bibliothéque des Textes Philosophiques, 1972, X2 (11736 15), p. 487.

first two conditions, the observed necessary extra heat production is produced by shivering and increased general activity and, after acclimation, by non-shivering thermogenesis of the brown adipose tissue. Only the increase in feeding and/or fat utilization can balance this supplementary energy expenditure. In rats, dogs and humans and other species, a total regulation by food intake compatible with a maintained body weight is observed (Johnson & Cark, 1947; Brobeck, 1948; Cottle & Carlson, 1954; Durrer & Hannon, 1962). In rats placed experimentally in cold ambient temperature (5 °C), the increased intake occurs readily. It is achieved initially by an increase in meal number and the extension of feeding to daytime (Kissileff, 1968), and later by an increase in meal size (Portet, 1981).

During exposure to cold, rats and mice increase their energy intake by selecting carbohydrates and not fats, as observed when energy imbalance has been produced by fasting (Andik & Bank, 1954). Some experiments have provided evidence that cold as a thermal stimulus of the skin can induce an increase in the urge to eat before inducing an energy loss. This response could be categorized as an anticipatory neurogenic response comparable to other neurogenic responses, such as pulmonary hyperventilation under a passive mobilization (Kraly & Blass, 1976).

In a warm ambient temperature, the temperature gradient becomes too small for the dispersion of the obligatory heat production needed for the maintenance of homeothermia. Because this basal heat production is obligatory, it is not reduced during acute or chronic exposure to a warm environment and rather increases. But food intake in rats and in humans is drastically reduced. At 33 °C rats fed *ad libitum* eat only 25% of their intake at 17 °C (Brobeck, 1948). In this condition, rat body temperature increases and they lose weight (about 2 g per day). In the heat, in rats and also in man, hypoactivity is observed. It is suggested that reductions in both food intake and activity are involved in the defence of homeothermia by restraining the part of heat production due to heat increment of food and to exercise. In this case, the fever and elevation of skin temperature could be the triggering stimulus for the reduced intake which would overcome normal stimuli in an emergency situation.

Exercise

Increase in muscular activity which can augment the basal metabolism intermittently or persistently two- or three-fold is another case of thermogenesis-induced feeding. Acute and chronic exercising induces lipolysis and therefore weight loss and the depletion of fat store.

Contrary to the situation with cold, such exercise is associated with considerable residual heat production, which has to be dispersed to avoid an increase of body temperature. This does not avoid an elevation of skin temperature, and creates a combination of contradictory signals which can explain the variance of feeding responses to exercise.

In rats, dogs and humans (Nikoletseas, 1980; Applegate *et al.*, 1982) convergent results indicate that forced acute exercise is followed immediately by hypophagia. In rats, the reduced food intake lasts 18 h after a bout of exercise in a runwheel or by swimming (Stevenson *et al.*, 1966). Weight loss resulting from exercise and augmented by the post-exercise reduction of intake is restored during the following days and eventually augmented. However, the observed augmentation of intake during this period does not exceed the deficit due to the preceding hypophagia. Thus, as after a prolonged fast, an acute elevation of energy expenditure by exercise is compensated for by reduced activity and/or by an increase in efficiency of food metabolism. This is indeed observed in rats (Mayer *et al.*, 1954; Thomas & Miller, 1958). However, in various species, a chronic moderate exercise, in addition to the basal activity and other energy expenditures, is normally balanced by an energy intake compatible with a maintenance of body weight.

In man, rest, sleep and increases of food intake alternate and are combined, following chronic exercise, to maintain or to restore the normal body weight.

Pregnancy and lactation

During pregnancy, a female's endogenous energy retention is associated with hyperphagia. This increased intake is a true hyperphagia since it exceeds the energy retention due to the gestation and leads to an increase in body fat of the pregnant female. This gravidic lipogenesis has been interpreted as a storage of energy anticipating the *post-partum* lactation. Increased food intake is also observed in lactating females (Wang, 1923; Kristal & Wampler, 1973; Fleming, 1976*a*, *b*).

3 Systemic and sensory stimulation to eat

The initiation of eating results from a combination of a systemic or metabolic signal and the oro-pharyngeal sensory activity of the food.

A given state of energy deficit assessed by a duration of food deprivation does or does not stimulate eating a food, depending on the sensory properties of that food. This trivial observation, in animals as well as humans, has given rise to the essential notion that the onset of eating and the initial strength of the engaged motor pattern of ingestion is a response to, and therefore is mediated by, the sensory activity of foods on oro-pharyngeal sensory receptors. This response is 'permitted' by a state of the central nervous system (CNS) induced and influenced by a metabolic stimulus related to energy imbalance, a state which is designated 'hunger arousal'. The food-specific response at a given level of the metabolic permissive action is designated 'the palatability' of this particular food at the time of testing. Palatability of food, thus, is not (as is often believed) the sensory properties of a food but the ingestive response to these sensory properties, according to the state of the systemic stimulation. It is synonymous with '*sensory stimulation to eat*' or with '*sensory-specific appetite*'.

This last word 'appetite' needs another preliminary qualification. Hunger, in addition to the subjective sensation experienced by man, can be used to designate the above-mentioned state of the CNS that results from a metabolic signal which has itself to be identified. The terms 'partial' or 'specific' hunger are without meaning. Appetite is the hunger-stimulated response to a particular food. The term 'specific appetite', used to designate a response adjusted to a specific metabolic need of a particular nutrient, is equivocal. It will be seen later that all appetites are sensory-specific and are simultaneously either caloric appetites or nutrient-specific appetites, or both.

The systemic or metabolic stimulus to eat or 'hunger signal'

In order to identify the nature of the metabolic state which generates the signal for ingestive responses to food as an external sensory stimulus, a first step is to examine this state in an animal model offered a constant familiar food. The variable of food palatability being eliminated, the effective stimulus to eat can then be investigated by two complementary methods. In the first, one can examine the temporal correlation of meal initiation with the nutritional status and particularly with the levels of energy metabolites in the blood. In the second method, the suspected metabolic parameter can be experimentally manipulated and the degree of change in feeding response can be examined.

Temporal correlation

In the first approach, three conditions of feeding may be studied: *ad libitum* feeding; feeding following a short-term food deprivation; and feeding on a fixed schedule of food access.

Ad libitum *feeding*

In the preceding chapter, it was shown that free-fed rats start a meal at a time which is dependent on the amount eaten in the previous meal. This suggested that a meal is initiated after a time dependent on the rate of the metabolic utilization of the previous meal (or store) and therefore on a given level of exhaustion of this store. This implies that a constant state related to the depletion of the store was present at the start of meals. However, to confirm the hypothesis it was necessary to see whether the same triggering signal was generated during the daytime, after a post-prandial interval more than twice as long as that occurring during the night, by demonstrating a difference in the rate of food utilization during the two periods.

A systematic relationship between a change in metabolic parameters and the times of meal initiation was only revealed by following the evolution of various parameters from meal to meal in free-fed rats. It was achieved by the difficult technique of intermittent or continuous withdrawal of blood via chronically implanted catheters in unrestrained rats (see Chapter 1).

The level of circulating free fatty acids was excluded as a possible triggering signal for eating. Plasma free fatty acids (PFFA) generally tend to increase in the pre-prandial period but their level at the start of nocturnal and daytime meals differs greatly (Steffens, 1969*b*; Le Magnen, 1976). During daytime, a high PFFA level is evidence for the involvement of lipolysis as a metabolic condition which delays meal onsets (Le Magnen & Devos, 1970). However, nothing indicates that this PFFA level just prior to the start of a meal acts as the triggering signal. A change of amino-acidaemia or in blood amino-acid pattern has never been demonstrated to be contingent on meal feeding. Data obtained by Melinkoff *et al.* (1959) relating changes in the blood spectrum of amino-acids to food intake in man have not been confirmed.

Metabolic parameters other than the level of substrates of energy metabolism in the blood have been suggested to be the hunger signal. The fact that the level of O_2 consumption and degree of body or skin temperatures are not significant as short-term stimuli will be discussed later in relation to the untenable thermostatic or energostatic theories of feeding. Oxygen consumption increases at meal initiation, presumably as an effect of pre-feeding activity (Le Magnen, 1976). The difference in responses to fasting for 12 h in the dark and 12 h in the light is not related to the difference between the fasting-induced fall of O_2 consumption during either of the two periods (Le Magnen *et al.*, 1980*a*). Hypoxia is not a stimulus to eat (Ettinger & Staddon, 1982).

There is now a considerable body of evidence for a role of glucose homeostasis and rate of glucose utilization in generating the hunger signal. It turns out that the feeding mechanism is linked to the complex mechanism which controls the supply of glucose to the tissues and maintains a constant blood glucose level (BGL).

The first evidence came from observing changes in BGL and gluco-regulatory hormone levels from meal to meal in *ad libitum* fed rats. Blood sampling every 10 min from one meal to the next yields the following results (Steffens, 1970; Strubbe & Steffens, 1977; Strubbe *et al.*, 1977). BGL was perfectly constant from the end of the prandial hyperglycaemia to the start of the next meal. In the last sample, which was drawn on average 5 min before the start of a meal, the BGL did not differ from other preceding specimens. Insulinaemia, peaked 15 min after the start of a meal, declined progressively during the post-prandial interval, and reached its lowest level in the last 5 min before the onset of the next meal. The pre-prandial insulin level was lower during the day than during the night. Insulin was excluded as a triggering factor, however, because an infusion-induced state of hyperglycaemia and its consequent endogenous insulin release did not delay the start of the subsequent meal. Finally, simultaneous portal and jugular blood sampling showed that the BGL in the portal vein was higher than that in the general circulation before, during and after meals and became identical with it only after 22 h of fasting.

Further experiments yielded different results regarding the pre-meal BGL (Louis-Sylvestre & Le Magnen, 1980*a*). A more sensitive technique was used to achieve a continuous withdrawal of a minute flow of blood (15 μl/min) through an intrajugular catheter for 10–12 h. Blood from a donor rat was restituted at the same flow rate through a femoral catheter. The continuous parallel measurement of the BGL (by an autoanalyser) and of the spontaneous meal pattern revealed several conditions.

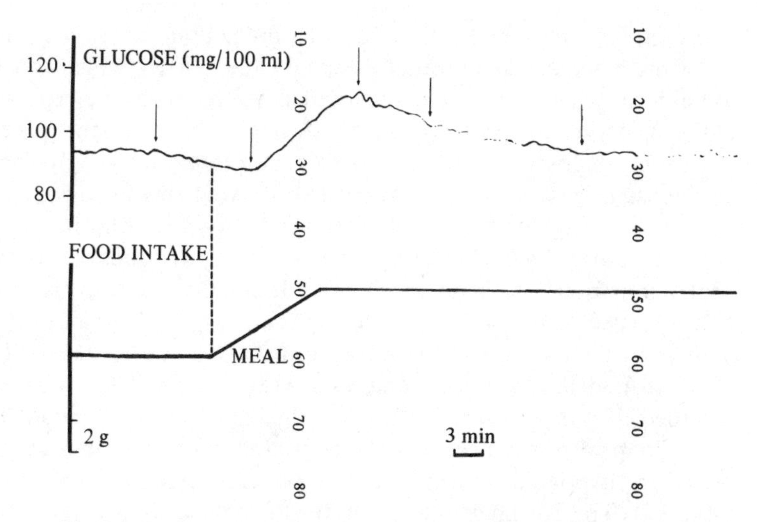

Fig. 3.1. Continuous recording of blood glucose level in free-moving rats. Onset of the meal is immediately preceded by a slight fall in blood glucose level.

(*a*) The start of each meal was always preceded by a fall of 6–8% in the BGL starting 6–7 min before eating began (Fig. 3.1). The deflection of the steady blood glucose line on the screen of the autoanalyser permitted the experimenters to predict with 100% certainty that the rat would go to its food-cup a few minutes later and begin to eat.

(*b*) The BGL continued to decline during the first minutes of the meal, before the post-absorptive sustained increase and the subsequent decline.

(*c*) Until the decline preceding the next meal, the BGL was constant and presented no visible oscillation. This regulated BGL was the same during the short nocturnal meal-to-meal intervals as it was in the long intervals during the day.

This decisive finding of a prandial fall of BGL has been replicated and fully confirmed by Campfield *et al.* (1984).

Thus, one could assume that an event related to a slight hypoglycaemia was the triggering factor for meal onset in the *ad libitum* condition. The increase in glucagon and the decrease in insulin release (plausible but not controlled effects of hypoglycaemia) could not be postulated as possible triggering agents. It was thus tentatively suggested that a sudden reduction in the blood glucose supply to specific chemosensors (later identified in the brain; see Chapter 8) gave rise to the behavioural arousal.

Preventing a meal by food withdrawal, for example at the beginning of the night, induces an immediate and rapid fall in BGL which appears to be the continuation of the fall recorded before and at the beginning of the meal when it is presented (Larue-Achagiotis & Le Magnen, 1982). Other data support the notion that rats eat to prevent hypoglycaemia and consume just as much as is needed to avoid this hypoglycaemia without bringing about other counter-regulatory mechanisms.

It is supported by the following experiment. Eight groups of rats were deprived of food for 3 h at various times during the circadian cycle: the first 3 h, then the second 3 h of the night and so on. The fall in BGL induced by this 3 h deprivation was recorded at 8 different times of day. The 3 h food intake of the same rats when fed *ad libitum* was also recorded. It was shown that there is a significant inter-individual and inter-group correlation between the normal eating rate during each 3 h period and hypoglycaemia induced by food removal for the same period (Fig. 3.2). The 3 h high intake during the night is associated with a relatively large fall in BGL during a 3 h fasting. During daytime a very low eating rate during the first 6 h was related to the absence of hypoglycaemia under fasting during the same hours. During the last 3 h of the day, the eating rate of six rats was highly correlated ($r = 0.90$) to their individual hypoglycaemia after 3 h of deprivation at this time (Larue-Achagiotis & Le Magnen, 1985*b*).

Other evidence for a critical role of blood glucose regulation and of glucoregulatory hormones in meal onset has been provided by experiments on intravenous self-administration of glucose, insulin and glucagon in free-fed rats (Jouhaneau & Le Magnen, 1980). Thus, strong evidence now exists for the presence, some minutes before the start of every meal in free-fed rats, of the beginning of a deficit in the glucose supply to the tissues, manifested by a minute decrease in the regulated level of glycaemia. What is the origin of this event in rats and the time of its occurrence after a meal (about 100 min at night and several hours during the daytime)?

The above-mentioned correlation between meal size and post-meal interval duration suggests that eating occurs as a result of a depletion of the gastrointestinal store loaded by a meal and of a consequent decrease of the inflow of fuel to the liver. Depletion of the hepatic glycogen store has been excluded. The glycogen content of the liver is far lower than the sum of metabolizable energy utilized from meal to meal. A comparison of glycogen content at the start of meals and 60 min afterwards during the night has shown that glycogen concentration in the liver does not decrease during the pre-meal time and rather increases from meal to meal throughout the night (Le Magnen & Devos, 1980*b*; Langhans *et al.*, 1982).

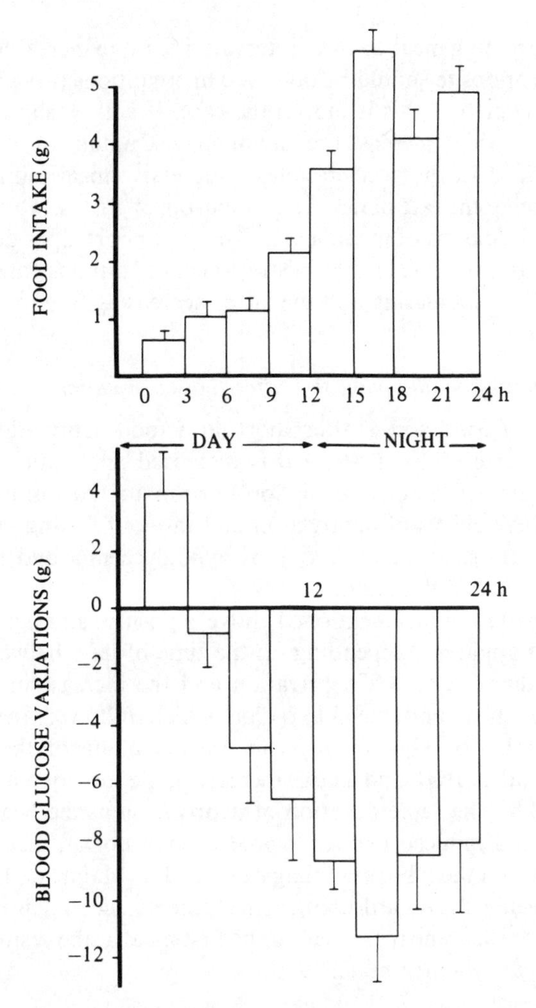

Fig. 3.2. During successive periods of 3 h throughout the day the eating of rats is correlated to the level of hypoglycaemia during each 3 h period when food is removed. This suggests that rats eat throughout the day at the rate needed just to maintain their blood glucose level.

If the depletion of the gastrointestinal store upstream to the liver is involved, the delay of onset of a meal is, for any particular amount eaten in the previous meal, dependent on the rate of metabolic utilization of the food. This rate can be increased either by an elevation of energy expenditure in lean tissues (i) or by fat synthesis (ii). It can also be reduced by a fall in energy metabolism (iii) or by substitution of endogenous lipid and glycerol for the exogenous fuel (iv). Under conditions (i) and (ii), an increase of cumulative intake should be achieved by shortening meal-to-meal intervals, i.e. by an increased meal frequency. The opposite should be observed in conditions (iii) and (iv). It will be seen later that this is indeed the case. It will be shown that all the conditions which increase the rate of glucose uptake by lean and adipose tissue lead to the acceleration of successive meal initiations, while a reduction of the rate of uptake or a sparing of glucose utilization will lead to a retardation of meal onsets. This is supported, as detailed below by evidence from insulin administration or continuous parenteral feeding, the former increasing and the latter decreasing, the frequency of meal initiation.

Meal initiation and meal size after food deprivation

On restoration of food access after short-term food deprivation (see Chapter 2), the size of the first meal is increased and initiations of subsequent meals are accelerated in proportion to the duration of deprivation. These effects of deprivation and those of fasting-induced hypoglycaemia argue again in favour of hypoglycaemia and related events as systemic stimuli to eat.

In the eight groups of rats mentioned above, a positive and significant correlation was apparent, depending on the time of day, between the fall in BGL induced by a 3 h deprivation and the increase in intake during the first meal and the 3 h period which followed the food removal. Three hours of deprivation at the beginning of the night induced a large fall in BGL and a large increase in the subsequent intake over the next 3 h. The same duration of deprivation at the beginning of the light period induced neither hypoglycaemia nor an increase in the subsequent intake (Larue-Achagiotis & Le Magnen, 1985*b*). Throughout the day, the same decrement in glycaemia under deprivation was correlated to the same increment in the first meal at the restoration of food access (Le Magnen *et al.*, 1980*a*).

Systemic stimulus of meal initiation under a feeding schedule

On a feeding schedule of one or several meals per day at fixed hours, meal size is not dependent on the regular interval of non-access to food which precedes the habitual meal. On the contrary, evidence exists that the size of the habitual meal anticipates expenditures to be made during the regular and fixed post-meal interval. This suggests that a condition identical to that observed in *ad libitum* rats is achieved at the regular time of food presentation and is involved in triggering the start of eating.

A series of elegant experiments by various investigators has demonstrated that a pre-meal insulin release and induced hypoglycaemia conditioned to the time of the meal and to food related stimuli is the basis for a conditioned hunger.

External stimuli associated with feeding can be shown to act as conditioned stimuli to eat or to reinforce eating. Rats exposed to their habitual feeding environment at 0, 30, or 60 min before their single daily meal, ate more during that meal. The effect of this exposure to food-related and previously experienced stimuli lasted 30 min (Valle, 1968). In another experiment, rats were fed two regularly spaced meals per day. For 11 days, meal presentations were associated with an irrelevant stimulus (sound or light). Another sound was delivered between meals. The rats were then transferred to the *ad libitum* condition and tested. The delivery of the conditioned stimulus, even at the end of a spontaneous meal, initiated meals. These conditioned meals could reach 21% of the daily free intake. However, this conditioned intake was compensated for by a reduced free intake during the day (Weingarten *et al.*, 1983).

It has been demonstrated that such feeding-associated stimuli are conditioned stimuli to eat, because they are conditioned stimuli of insulin release and therefore hypoglycaemia. After a repeated association between a visual or gustatory stimulation with tube-feeding in the stomach, the external stimuli induced a conditioned hypoglycaemia (Deutsch, 1974). The same effect was obtained by association of an olfactory stimulus with the repeated injection of insulin or with an intravenous or intraperitoneal glucose load (Woods, 1976). Even the procedure of injection (saline injection) induced the increase of plasma insulin concentration and consequently hypoglycaemia. Furthermore, after repetition of feeding initiated by insulin injection, injection of saline as a conditioned stimulus initiated feeding. Following a repeat retardation of feeding by injected glucagon, injecting saline delayed the initiation of a meal (Balagura, 1968; Balagura *et al.*, 1975).

The involvement of this conditioned change in blood glucose in triggering meal onset in a feeding schedule was finally demonstrated directly (Woods *et al.*, 1977). Rats received their meal at fixed hours twice a day. At the habitual time of the meal and of its expectancy by rats an increase in insulin concentration was observed. An external olfactory stimulus regularly associated with the meal had the same effect. The time of day became the conditioned stimulus for insulin release only with regular meals. The best stimulation was obtained by a combination of the time and the associated odour.

When an odour has become a conditioned stimulus for insulin release, the addition of the odorant to food as a flavour enhances the palatability of that food and the amount eaten (J. Louis-Sylvestre, A. M. Reynier-Rebuffel and J. Le Magnen, unpublished).

In man

In human subjects eating their meals at regular hours, what is the metabolic state present prior to the habitual meal? According to some authors, glycaemia would be constant between meals without any indication of a pre-meal fall (e.g. Bernstein & Grossman, 1956).

It will be seen below that in man as in rats, external food stimuli and presumably the time the meal is given act to induce pre-meal insulin release as a reflex. Omitting a meal produces hypoglycaemia. Even the omission of the breakfast (the continental type) produces a significant hypoglycaemia by lunch-time and an increase in the amount eaten and in eating rates during this lunch (Bellisle *et al.*, 1984). The hunger rating, in deprived subjects, was found to be negatively correlated to the BGL (Janowitz & Ivy, 1949; Fryer *et al.*, 1955; Knight *et al.*, 1980).

Effects on meal initiation of manipulating the systemic stimulus

While the same familiar food is still used and therefore remains a constant sensory stimulation to eat, the putative systemic stimulus which permits this sensory mediation of eating can be modified by various means.

Exogenous insulin

The injection of a single dose of regular insulin to rats elicits a dose-related augmentation of the subsequent food intake over some 4–6 h (Booth, 1968). This feeding response to exogenous insulin occurs after a latency of 30–40 min. During this latency, feeding is inhibited (Steffens, 1970). The 4–6 h augmented intake is achieved by an increase in meal frequency rather than by increased meal size (Steffens, 1970). The intake-promoting effect is different during the two parts of the diurnal cycle. Under continuous infusion of 0.1–0.6 IU/h, a dose-related elevation of the 12 h intake is observed during the day, reaching twice as much (or more) as the intake of control rats. During the night, the same infusions only produce a slight (25%) dose-independent elevation (Larue-Achagiotis & Le Magnen, 1979). A gastric or intravenous glucose load prevents the insulin-induced hypoglycaemia and the subsequent overeating (Booth & Pitt, 1968).

The increases in intake following injections of a single dose and 12 h infusions of regular insulin are compensated for by a subsequent reduction in intake. The body weight is thus not affected. On the contrary, daily injections of long-acting insulin (protamine-zinc insulin, PZI) or 24 h infusions of regular insulin produce a permanent overeating

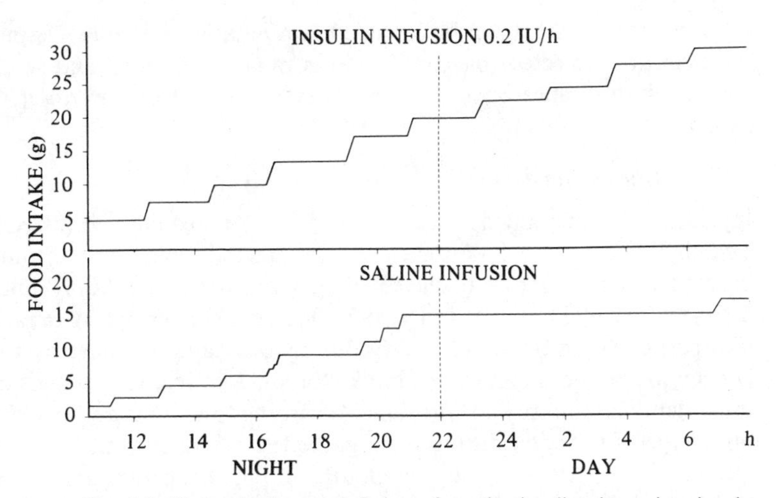

Fig. 3.3. Under continuous infusion of regular insulin via an intrajugular catheter, the daily increase of food intake by rats is achieved through a moderate increase at night and a considerable increase during the daytime. The latter is mainly due to an increase in meal number.

and therefore a progressive weight gain until obesity is reached (MacKay *et al.*, 1940; Hoebel & Teitelbaum, 1966; May & Beaton, 1968; Panksepp *et al.*, 1975) (Fig. 3.3). This insulin-induced hyperphagia is achieved by an increased meal frequency and the suppression or abolition of the normal daytime hypophagia (Larue-Achagiotis & Le Magnen, 1985*a*).

In man, a single injection of regular insulin also elicits feeding and a severe hunger. As in rats, both occur after a latency of 30–40 min (Janowitz & Ivy, 1949).

The BGL drops readily following insulin injection. In rats the high dose (10 to 20 IU/kg) used to induce feeding produces a fall in BGL to 50–60 mg/100 ml without delay. This insulin-induced hypoglycaemia is due to a massive glucose uptake by the tissues, promoted by insulin. In this condition of high tissue uptake of glucose and therefore of severe hypoglycaemia, which is present during the 20–40 min latency to eat, feeding is not stimulated. Thus, hypoglycaemia *per se* is not the systemic stimulus to eat. When rats and humans begin to eat, blood glucose is still at a very low level (Janowitz & Ivy, 1949; Booth & Pitt, 1968). But the fact that infusing glucose at this time prevents eating suggests that feeding is then stimulated in a state of tissue glucopenia subsequent to the initial high glucose uptake. It will be seen later that this glucopenia which appears to be the effective stimulus to eat is brain tissue glucopenia.

Insulin *per se* can again be excluded as a putative stimulus. Despite the high dose injected, plasma insulin is rapidly degraded and so its concentration comes back to normal level long before the onset of eating.

Glucose analogues

Injections of three glucose antimetabolites – 2-deoxyglucose (2-DG), 3-methyl-glucose and 5-thio-glucose – peripherally and also centrally injected have been shown to elicit feeding in rats, mice, monkeys (Smith & Epstein, 1969; Houpt & Hance, 1971; Ritter & Slusser, 1980). In rats, intraperitoneal injection of 2-DG during the daytime shortens the latency to eat and increases the intake for 4–5 h in free-fed animals or those deprived for only a short time. According to various authors, the same type of injection during the night is less active or without effect (Naito *et al.*, 1973; Jones & Booth, 1975; Larue-Achagiotis & Le Magnen, 1979; Nagai *et al.*, 1982). Glucose infusion blocks the effect of 2-DG just as it blocks the effect of insulin (Stricker & Rowland, 1978).

Injecting glucose analogues produces a severe state of tissue glucopenia and the mobilization of a series of counteractions to this emergency condition: adrenergic stimulation of lipolysis; adrenal catecholamine release and a resulting hypoinsulinism; and stimulation of glucagon release and of hepatic glucose production. Hyperglycaemia results from this metabolic pattern which mimics acute diabetes. Hyperglycaemia in this condition seems to antagonize the feeding response. After adrenalectomy, the hyperglycaemia response to 2-DG is abolished and the feeding response is augmented (Frohman *et al.*, 1973; Jones & Booth, 1975; Thompson & Campbell, 1978). It will be seen later that 2-DG, unlike insulin, acts directly on brain sites to stimulate feeding and that this action is independent of its promoting effect on lipolysis and hyperglycaemia (Le Magnen, 1983). Whether or not this action of 2-DG reproduces the physiological hunger stimulus will be discussed below. Already it is possible to conclude that the severe tissue glucopenia induced by 2-DG as an emergency condition initiates other mechanisms in which the normal stimulation of eating is upset and participates therefore only modestly in the relief of the metabolic stress.

Diabetes

Rats made diabetic by alloxan or streptozotocin treatment exhibit the classical symptoms of insulin-dependent diabetes – glucose intolerance, hyperglycaemia and hypoinsulinism. Initially and throughout the next 10 to 12 days, they are hypophagic and later become hyperphagic when offered a carbohydrate-rich diet. During the initial hypophagia, a high level of circulating free fatty acids and weight loss indicate lipolysis. This hypophagia, then, is similar to that observed in other cases of energy metabolism where lipidic fuel is substituted for carbohydrates. The other cases are, as will be discussed in Chapter 6, the normal daytime condition, which is effectively analogous to a physiological acute diabetes and the condition which follows a forced weight gain (see Chapter 6). Hypophagia in diabetic rats is replaced by hyperphagia and symptoms of the initial lipolysis also disappear (Carpenter & Grossman, 1983*a*).

Enteral and parenteral feeding

Short- and long-term injections or infusions of free fatty acids or of amino-acids do not have significant or physiologically relevant effects on free food intake (Booth & Campbell, 1975; Carpenter & Grossman, 1983*b*).

Results generally agree that injecting a glucose solution during or at the end of a meal-to-meal interval in free-fed rats does not delay the meal onset despite the induced hyperglycaemia (Bernstein & Grossman, 1956; Steffens, 1969*c*, 1970). In fact, injecting 20 mg of glucose intravenously (which suppresses only the occurring hypoglycaemia) at the predicted time of onset of a meal, does indeed retard the meal (Campfield *et al.*, 1984).

Similarly, hyperglycaemia, and therefore the abolition of hypoglycaemia by glucagon (Balagura *et al.*, 1975; Geary *et al.*, 1981) and by adrenaline injections (M. Lagaillarde and J. Louis-Sylvestre, unpublished), just before the rats' meal, effectively retard the start of this meal. The retarding action of these hormones on meal onset argues strongly in favour of the critical role of the pre-prandial hypoglycaemia and related events.

Finally, chronic enteral and parenteral feeding and the compensatory reduction of oral intake provide some information about the systemic stimuli to eat or not to eat. In many experiments in rats, dogs and other species, it has been demonstrated that single or continuous intragastric, intraduodenal or intravenous loads of nutrients are followed or accompanied by a reduction in oral intake which partly or entirely compensates for the caloric supply. Experiments in which nutritive loads are given long before an oral meal and act on the subsequent or current feeding pattern (Share *et al.*, 1952; Janowitz & Hollander, 1955) have to be distinguished from other experiments in which only the effect of the load on the size of a meal is investigated.

A single intragastric administration of a percentage of the daily previous intake is followed by a compensatory reduction of the oral intake over the subsequent 6 h. The reduction of intake is realized mainly by prolongations of meal-to-meal intervals (Booth, 1972*a*). A continuous gastric tube-feeding over 24 h by the same amount of the liquid food as previously eaten may lead to a total abolition of oral intake. When food intake is only reduced, this reduction is again realized through a reduced meal frequency, indicating a suppression of the stimulus which normally triggers food intake (Thomas & Mayer, 1968; Thomas & Mayer, 1978). Intestinal chronic infusion gives the same result (Glick & Modan, 1977). Continuous intravenous feeding through a chronic intrajugular catheter is also compensated for by a reduced oral intake (Nicolaïdis & Rowland, 1976). Only carbohydrate solutions are effective. When the infused carbohydrate solution provides 100% of the previous intake, the compensation was only partial. The residual intake was nocturnal, with a meal pattern identical with that normally observed in the daytime. The reduction of oral intake was augmented by adding insulin to the perfusate (Rowland *et al.*, 1975). This argues again in favour of a deficit in glucose uptake by the tissues as the stimulus to eat. It is well known that, compared to oral glucose load, intravenous infusion is a poor stimulus of insulin release. Despite the previous supply of calories by the glucose infusion, the compensation is imperfect, presumably because the tissue glucose uptake remains at a low level as a result of a deficiency in insulin release. Adding exogenous insulin abolishes the residual stimulus to eat.

Sensory-specific stimulation of eating

What happens when the palatability of the above constant and familiar food is changed? Adding a non-caloric sweetener or greasy material to that food augments intake in a given meal in free-fed rats or those deprived for a short time, and in the former it also increases the cumulative 24 h intake. From this observation it may be said that the food has been made more 'palatable'. Adding a bitter-tasting compound reduces the same feeding responses – the food has been made less palatable. In the *ad libitum* condition the recording of meal pattern shows that the augmented intake of the more palatable food is achieved by increases in both meal size and meal frequency while the reduction in intake of the less palatable version is achieved by smaller and less frequent meals (J. Le Magnen, unpublished). Thus, in an identical state of systemic stimulation, a graded and sensory-specific stimulation of meal initiation and meal size is dependent on the orosensory activity of food. We shall come back later to the origin of this differential palatability of foods and to its relation with their nutritive properties.

This differential palatability is a continuum. We have no reason to distinguish positive and negative palatabilities. These notions close to the ideas of preference and aversion are misleading, if it is not made clear that positive and negative palatability, preferences and relative aversions are so called by reference to a particular food control. The difficulty and misunderstanding arises from the fact that the referred food to which the palatability of other foods is compared is not defined. Thus, a food claimed aversive and in the range of negative palatability compared to one food, will be claimed 'preferred' and in the range of positive palatability when compared to another food. Using water as a standard for fluid intake partly avoids this confusion.

Evidence for this continuum and for the additive or synergistic effect of internal and external stimulation of eating has been accumulating for a long time. In rats, intake of a food under increasing durations of food deprivation is blocked or reduced by the introduction of increasing concentrations of a bitter-tasting compound into the food (Miller, 1955; Williams & Campbell, 1961). It will be seen later that lateral hypothalamic (LH) electrical stimulation mimics hunger and stimulates feeding (Tenen & Miller, 1964). The intensity of the stimulation needed to stimulate intake of a food increases when its palatability is lowered by adding quinine. A palatability of zero would be that of a material such as sand or stone which would be refused even under maximal hunger or systemic stimulation. A maximal or infinite palatability would be that of a food accepted and eaten theoretically in the absence of an internal signal. This last condition is difficult to conceive and has not been experimentally ascertained. True aversion must be distinguished from unpalatability. True aversion is a defensive reaction implying nausea and eventually vomiting. Brain mechanisms other than those involved in the continuum of palatability are involved in this defensive response to a food which is spontaneously, or has been rendered, offensive.

However, eating a food in a state of satiety induced by the immediately preceding intake of a sensorily different food is a trivial experience within human meals. The first experimental study of this phenomenon in an animal model was performed in the following way. Rats were habituated for 32 consecutive days to a 2 h morning meal during which they were offered at random the same synthetic diet flavoured by adding an odorant, *A*, *B*, *C* or *D*, to the food. After some days of intake varying with the marker, the 2 h daily intake of the diet stabilized and was identical whatever flavoured version was offered. Then, every two days, the 2 h meal consisted of the successive presentation of the four flavoured versions of the diet, at 30 min duration each. The order of presentation was varied at random in successive tests. In this successive

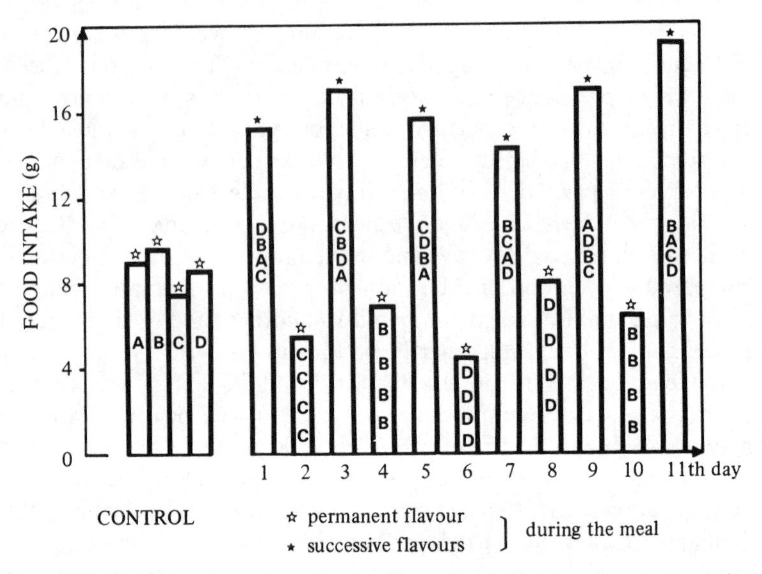

Fig. 3.4. Sensory variety of successively offered food increases the total size of the meal eaten by rats. After a long habituation, an equal amount of the same food is eaten during a 2 h meal regardless of its labelling by one of four added odorants (*A–D*). The 2 h intake is considerably elevated when the four flavoured forms of the diet are offered successively during the meal.

presentation, rats overeat considerably (Fig. 3.4). Whatever form of food was given first, the initial 30 min intake was close to that of the 2 h meal consumed by control animals. Nevertheless, rats ate again on being presented with the next form. But until satiation, intake decreased from the first to the last 30 min presentation (Fig. 3.5). It was almost nil for the last period. In a comparable experiment, three flavoured versions were presented *simultaneously* after habituation in a 2 h choice. In this situation, the total meal size was also transiently augmented though to a lesser degree than under conditions of successive stimulation (Le Magnen, 1956, 1960).

More than 20 years after the original publication, these experiments have been repeated and the results fully confirmed (Rolls *et al.*, 1983; Treit *et al.*, 1983).

Although the significance of this phenomenon and its mechanism will be commented upon in Chapter 4 as evidence for a 'sensory-specific satiation', it is mentioned here because it is also evidence for a 'sensory-specific stimulation of eating'. In the process of satiation, due to the progressive accumulation of food in the stomach, the initial palatability of the first-presented food is progressively abolished; but the subsequent intake of food with a new flavour proves both the

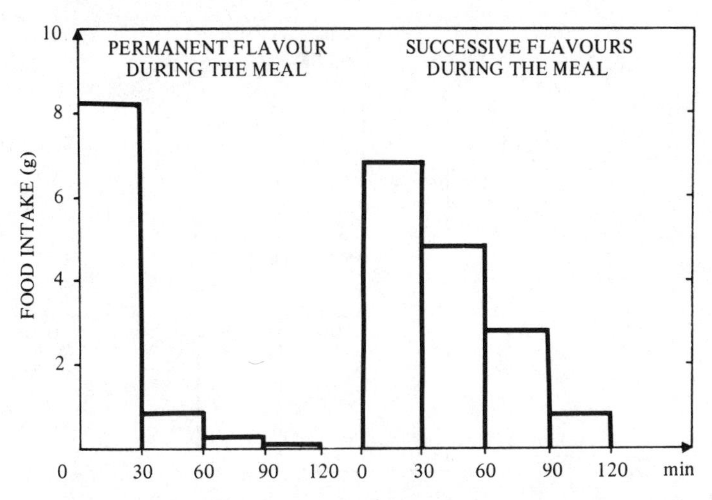

Fig. 3.5. Whatever the order of presentations of the four flavoured forms of food in Fig. 3.4, intake decreases during the four successive 30 min presentations. This indicates the role of gastric negative feedback adding its effects to the oral sensory-specific satiation.

sensory-specific stimulation of eating and that satiety reached with the preceding flavour is only 'partial'. The decreasing size of intake of the successive versions until total satiation is reached indicates that a remaining systemic stimulation is still active at the presentation of each new flavour until the final and total blockade by the gastrointestinal feedback. In man, the use of the sweet dessert, i.e. of the most palatable food at the end of an ordinary meal which then can be eaten as a result of a minimal residual hunger signal, is good evidence for this process. In addition, it has been shown that the shift from one food to another one during a varied meal in rats was accompanied by a new induced insulin release as a reflex (Louis-Sylvestre, 1983*b*). This may act to enhance the hunger-dependent palatability in a positive feedback opposing the negative one from the stomach.

The exact law (additivity or synergy, of the combination between palatability level and hunger arousal) assessed by duration of food deprivation has been very little investigated and is unknown. This gap in experimental studies of food intake is extremely surprising. However, a first approach has been carried out in man (Bellisle *et al.*, 1984).

The microstructure of a meal was studied by a recording of the chewing–swallowing pattern. During the meal offered at lunch-time, 4 h after a normal breakfast, subjects ate either a high or a low palatability food, determined by a previous subjective scaling. Meal size and meal

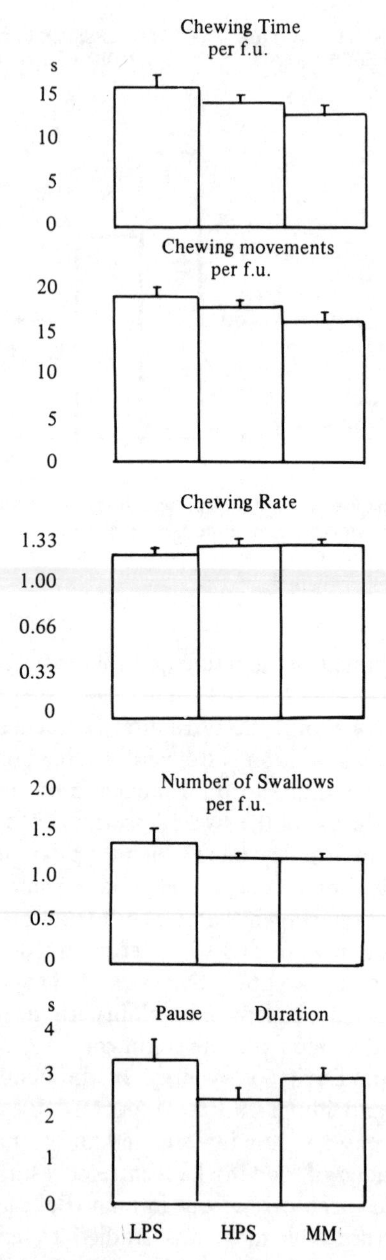

Fig. 3.6. Five parameters of the microstructure of meals in 10 lean human subjects. LPS = low palatability single-flavour meal; HPS = high palatability single-flavour meal; MM = mixed meal; f.u. = a standard food unit.

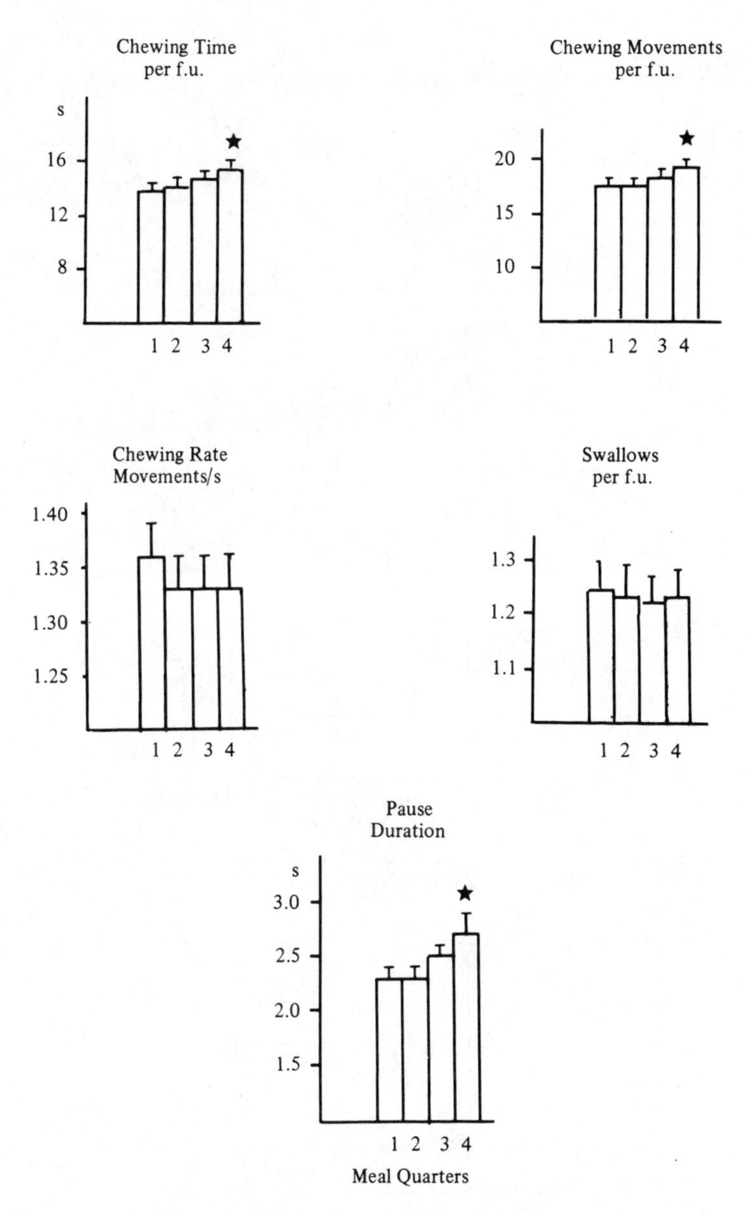

Fig. 3.7. Evolution of five parameters of the microstructure of meals between the first and the fourth quarters of meals in 10 lean human subjects.

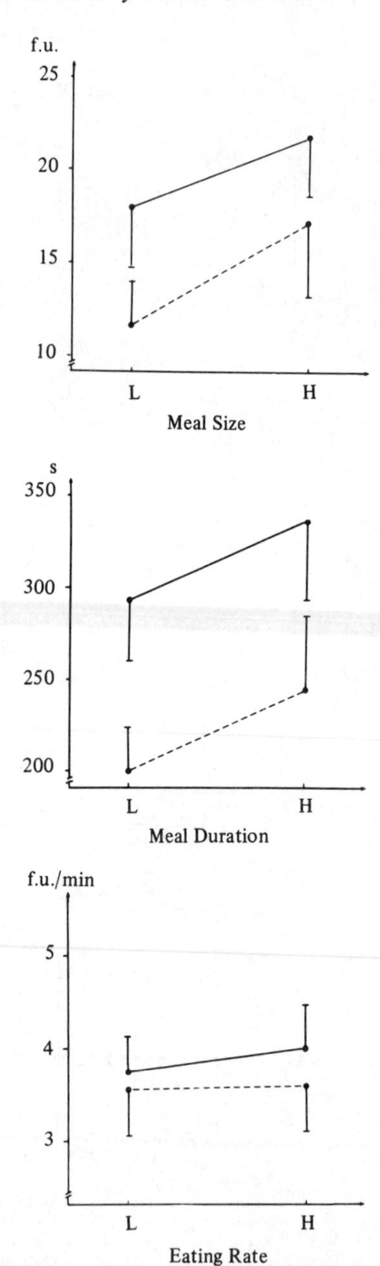

Fig. 3.8. Meal size (number of food units), meal duration and eating rate (number of food units ingested per minute) in 10 human subjects. L = low deprivation, H = high deprivation. The solid line represents a high-palatability condition and the dotted line a low palatability one.

duration and various parameters of the chewing–swallowing pattern at the beginning of the meal are quantitative measures of the different palatability levels of the two foods (Figs. 3.6 and 3.7). The increase of deprivation at the time of the meal (15 h instead of 4 h) augmented the characteristics of the palatability level of the two foods (Fig. 3.8). Eating the two foods after an increased deprivation has the same effect as the improvement of their palatability. The fact that in man hunger elevates the pleasantness of foods as a sensory stimulus (which is the subjective correlate of palatability) had already been observed by Aristotle more than 2000 years ago!

4 Determinants of meal size

John Brobeck (1955) was the first to point out that the cumulative intake of foods in a feeding pattern was the product of meal size and frequency and that a study of feeding must involve these two different parameters. However, the majority of studies devoted to feeding behaviour is limited to observation of 'eating' rather than of 'feeding'. An abundant literature has been focussed on the determinants of amount of food eaten in a bout of eating (or meal), from its initiation until its end. Many of these reports dealing with determinants of meal size are presented as studies on 'mechanisms of satiety'. This term has created great confusion. 'Satiety' is the state during which, from the end of one meal to the occurrence of the next, a subject does not eat, i.e. is not stimulated to eat. It is a passive state of no hunger. The termination of a meal, which finalizes meal size, is the onset of satiety; but the mechanism which leads to this termination during a meal is not 'satiety'; it is the process of 'satiation'. The onset of satiety is also the achievement of satiation.

When animals or humans are stimulated to eat and are offered a food, what determines the attainment of satiety after eating a given amount of that food? An apparently stupid answer is that in order to be satiated, it is necessary to eat: in other words, satiation is brought about by eating, i.e. by chewing and swallowing the food, by its accumulation in the stomach and its transit to the intestine and by eventual changes at post-absorptive level. All the various and successive actions of the food at these different levels (its sensory properties, its volume, its caloric content and caloric density and other nutritive properties) may be and indeed are involved.

The origin of food-specific palatability

The strength of the initial sensory stimulation to eat, as a positive feedback, is a determinant of meal size. What is the origin of this food-specific stimulation or palatability of foods? Activities of foods as stimuli of cephalic sensory receptors are independent of their nutritive activity. Thus, the initial response to an unfamiliar, i.e. unexperienced food, is irrespective of its nutritive properties.

Acceptance and rejection of a food and the amount eaten are

apparently randomly modified by adding odorants. However, when the neophobic response to a novel item is overcome, some genetically programmed responses to gustatory stimuli are in accordance with nutritive characteristics. Caloric sweeteners are palatable; bitter-tasting toxic foods are unpalatable; but non-caloric sweeteners are also palatable and non-toxic bitter-tasting agents unpalatable.

As a result of a process commonly designated 'familiarization', the differential palatability of food items becomes roughly adapted to their respective nutritive qualities. This raises the question of how the adjustment of food palatability to nutritive properties and to metabolic demands comes about.

It has been clearly demonstrated that the eating response to the sensory properties of foods is learned. Sensory activities of foods are state-dependent *conditioned stimuli to eat*. In a conditioning process, the post-ingestive nutritive activity of the food acts as an *unconditioned stimulus*.

The first historical evidence about the learning of a specific appetite was provided by Harris *et al.* (1933). In the 1950s came the demonstration that learning was also involved in caloric appetite, i.e. in palatability adjusted to the caloric properties of foods. In Europe, at this time, 'one trial and long delay' learning was acknowledged as characteristic of the conditioned taste aversion. In the ensuing years there has accumulated an enormous amount of literature (at least 1500 titles) exclusively devoted to this conditioning of a true aversion by a post-ingestive deleterious action of the orally tasted food. However, many experiments have provided evidence that the caloric supply by food also acts as an unconditioned stimulus to modulate food-specific palatability. This occurs at all levels of the palatability continuum.

The following two experiments are considered to be conclusive. (*a*) At two daily meals rats were presented with food labelled by the addition of one of two odorants, *A* or *B*. One group of rats eating form A and another eating form B were immediately injected intraperitoneally with a glucose solution adjusted to provide 25% of the caloric value of the preceding oral intake. For each group, the intake of the alternative version (*a* or *b*) of the diet was followed by a saline intraperitoneal injection. In both groups, the intake of the version not supplemented with glucose developed progressively. Finally, a larger amount of this version was preferred to the other one, as if it was of lower caloric density (Fig. 4.1) (Le Magnen, 1959*a*, *b*, 1960, 1969).

(*b*) In the other experiment, rats were presented with two starch solutions. One of a low caloric density was labelled by adding odorant *A*, and the other one of high caloric density by adding odorant *B*. After

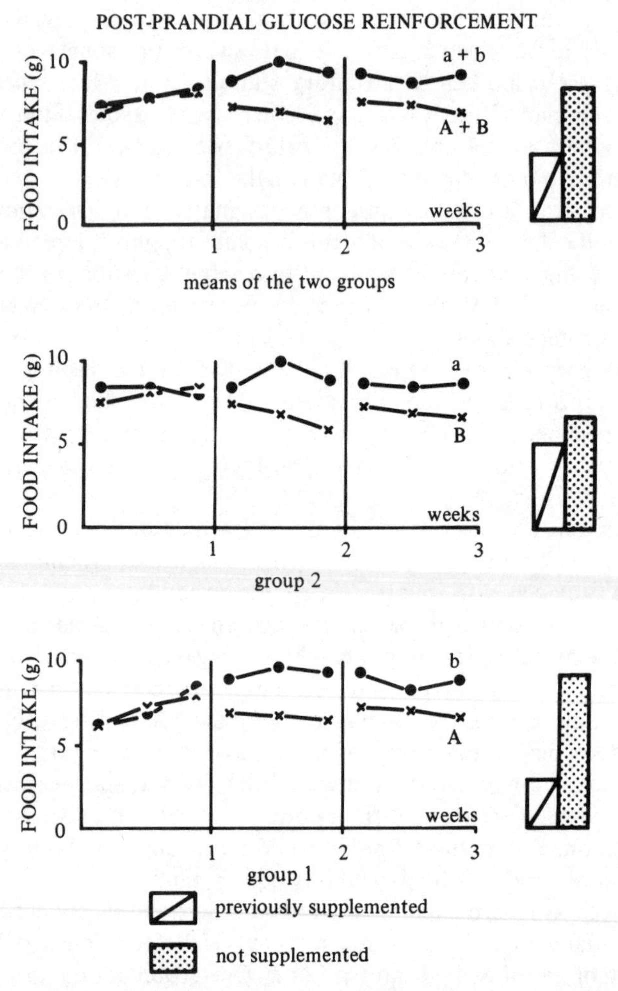

Fig. 4.1. Daily, rats are presented alternately with two flavoured forms of the same diet. The forms *A* in group 1 and *B* in group 2 and the intake of the other forms *b* in group 1 and *a* in group 2 are followed by a saline injection. In a final choice between the two flavoured forms, that which was previously supplemented with glucose is consumed less, indicating post-ingestive conditioning of palatability.

some days of alternate presentations of the two versions, a starch solution of an intermediary caloric density was substituted, with either *A* or *B* added. Rats drank a larger amount of the solution labelled by the odour associated previously with the low caloric content, either in a single-bottle presentation of each version or in a simultaneous choice (Booth, 1972*a*).

However, Booth (1972*c*) also noted that the conditioned differential eating response to the two versions in the above experiment was not due to a difference in the initial strength of stimulation. In a choice, the difference in intake was not due to the initial rate of drinking which was identical with both flavoured versions. It was due to a slow decline, until satiation, of the rate of intake on the preferred version. Later, he argued that the decline of the rate of intake as an oral negative feedback was also a conditioned response (Booth, 1972*c*, 1977).

Many other experiments have shown that conditioning was responsible for a difference in strength of the initial stimulation. Comparisons between intakes of a saccharin versus a sugar solution were instructive. Similar amounts of saccharin and sugar solutions were drunk when presented to rats for a short time, and when the two solutions had equal sweetness. The saccharin intake was repeatedly associated with the administration through gastric tubing of (*a*) an isotonic saline solution in one group of rats, and (*b*) a glucose solution in another group. In a final test, the saccharin intake was almost nil in the first group, but high in the second, despite the associated deliveries to the stomach being reversed (Capretta, 1964). In a similar experiment, two groups of fasted rats received either a sucrose or a saccharin solution for one daily meal. In the first group, rats increased their sugar intake daily over successive days while the rats of the second group, after a rise of intake on the first day, refused to drink the saccharin solution on the third day of fasting (Le Magnen, 1954).

The post-ingestive reinforcement has also been clearly indicated by the following experiment: a place or a flavour was associated with either the sham-feeding or real feeding of the same amount of food in rats previously deprived for 17 h. Preference for the location and the flavour developed only in rats which were fed normally (Van Vort & Smith, 1983). It is notable that, in contrast to humans, in rats an initial spontaneous unpalatability due to bitter-tasting compounds cannot be reversed, at least as assessed in a choice by association to a post-ingestive repletion (Stark, 1963; Rozin *et al.*, 1979). However, the initial aversion to ethanol solutions in ethanol-naive rats can be reversed in rats made dependent on ethanol (Le Magnen & Marfaing-Jallat, 1984).

What is the unconditioned stimulus in the post-ingestive action of the food? Various experiments show that the source of the reinforcement is either in the stomach or in the intestine, and/or at systemic level. At these various levels, the reinforcement is linked with satiation, i.e. with the recovery from hunger.

In one experiment with rats, the drinking of two non-nutritive solutions was associated with the administration by gastric tubing of either nutritive solution or a saline solution. A ligature of the pylorus prevented the stomach emptying. Rapidly, a preference was exhibited in a choice between the two flavoured solutions for the solution associated with the nutritive liquid diet. Thus food in the stomach alone may act as reinforcer (Deutsch & Wang, 1977). In another experiment, it was shown that only the first 5 min of intake of either of the two flavoured versions of different caloric density were necessary to induce the flavour-mediated preference for one of the two versions (Booth & Davis, 1973).

Many investigators have shown that recovery from illness or malaise acts as an unconditioned stimulus to establish oral preferences. The relief of vitamin deficiency induces preference for associated food items (Seward & Greathouse, 1973). The recovery from poisoning by LiCl acts as a positive reinforcer (Green & Garcia, 1971; Hasegawa, 1981). This and the above observations that the shift from hunger to satiety acts in a positive reinforcement of food preferences, support Hull's theory of reinforcement by drive reductions (Hull *et al.*, 1951).

Irrespective of the sources of signals and modes of their transmission to the brain, reinforcement is of course processed by the brain, where systems rewarding self-electrical stimulation (some of them at least) are presumably involved in the *food reward* (Chapter 8). The concept of food reward is somewhat equivocal. Indeed sensing and eating a food pellet allow the rat to learn and therefore reward an instrumental response (lever-pressing) to obtain the pellet; but the study of the rewarding by food of an instrumental response is of limited interest. Physiologically, food is the reward for learning an oral motor response – 'the eating response' – and it is not really useful to study this learning indirectly by substituting an instrumental response. In this learning of food palatability, the classical laws of transfer from a primary to a secondary reinforcement are in action. Because the post-ingestive action of the food is rewarding (primary reinforcement), the external sensory activity of the food also becomes 'rewarding' (secondary reinforcement). This sensory reward (manifested subjectively in man by pleasantness) supports the response acutely and may be involved in 'second-order conditioning'. The delivery of drops of a sweet solution reinforces the learning of an operant response level in satiated rats (Guttman, 1953).

In man, food preferences and aversions are profoundly affected by sociocultural contexts. Through social transmission and parental education, the young learn what is good or bad to eat much in the same way as they learn what they should or should not do. This social learning permits or prohibits the experiencing of foods. Some of them are and remain distasteful, even without being tasted, merely because they are prohibited by social beliefs or by scientific knowledge (Birch & Marlin, 1982). Other foods are experienced and individual likes and dislikes can be modulated by the post-ingestive conditioning (Pliner, 1982; Pliner *et al.*, 1985). Liking for coffee or another bitter-tasting beverage containing caffeine or for alcoholic beverages, even though they may taste awful initially, are obviously reinforced by the psychotropic and/or anxiolytic effects. A study of prisoners coming back from deportation camps in a tragic state of malnutrition and specific deficiencies has shown extreme changes of previous likes and dislikes. Some who drank their coffee unsweetened before deportation, liked it extremely sweet afterwards. Others had developed a craving for fruit, as sources of vitamins, which they disliked before deportation (Berger & Le Magnen, 1957). Elegant experiments similar to those performed on rats (see pp. 49–50) have provided evidence for conditioning of differential intake foods based upon their flavours and upon their previous associations with the caloric content, even when this is unknown to the subject. In the same series of experiments, it was shown that the 'pleasantness' of the food eaten is reduced when subjects eating the food were satiated and augmented when they were hungry (Booth *et al.*, 1976, 1982).

The oro-gastrointestinal process of satiation

Many experiments have tested the respective roles of food passing the mouth and of food filling the stomach and/or the intestine in the negative feedback which counteracts the initial stimulation to eat until satiety. In these experiments, procedures were developed to test separately the effect of the mouth and oro-pharyngeal cues alone, of the stomach or intestine alone, of mouth plus stomach, of mouth plus intestine alone, and finally of mouth plus stomach plus intestine.

Mouth

Sham-feeding in rats, dogs and monkeys demonstrates convincingly that the cumulative sensory action *per se* of food passing the mouth cannot induce satiety. Rats offered solid or liquid food in a condition of sham-feeding eat for several hours. They then terminate the enormous meals and, after a short interval, resume their sham-drinking or sham-feeding. From this, it has been argued that they maintain a meal pattern, which indicates the participation of oro-pharyngeal cues in satiation (Kraly *et al.*, 1978). However, these stops could be the effect of fatigue rather than of an established satiety. An important point is that the sham-feeding pattern may not be exhibited immediately.

Oesophagotomized rats presented with a concentrated sucrose solution maintain their previous drinking pattern during the first sessions of sham-feeding. The unlimited intake appears progressively (Mook *et al.*, 1983). By recording the licking rate of a liquid food, it was observed that it increased progressively during successive sessions of sham-feeding (Davis & Campbell, 1973). This suggests that, under these conditions, there is a satiating effect of foods passing the mouth and this is progressively extinguished. This extinction in turn indicates that the oro-pharyngeal satiating property of foods is a conditioned response which is reinforced by the post-oral action and disappears rapidly in its absence (Yokel & Wise, 1975; Schuster & Johanson, 1981). This argues in favour of a sensory and therefore food-specific satiating power (Booth, 1979). The increase in meal size in meals of varied content (described above) also argues in favour of this notion.

In sham-feeding, the persistant and increasing unreinforced response is reminiscent of the 'rebound' increase in operant responding during extinction following intravenous amphetamine or insulin self-administration or intra-cranial self-administration (Yokel & Wise, 1975). The latter is stopped by inflating a balloon in the stomach like sham-feeding does when food enters the stomach (see Chapter 8).

A temporal contingency between sham-feeding (and therefore pre-gastric stimuli) and the post-oral actions of the food is needed to re-establish a normal meal size. A maximal inhibition of sham-feeding is obtained when food is put into the stomach 12 min after the sham-feeding begins. At this time, the maximum inhibition is higher than when food is put into the stomach before or at the start of sham-feeding (Antin *et al.*, 1977; Kraly *et al.*, 1978; Kraly & Smith, 1978).

The opposite condition to sham-feeding is achieved by self-administered tube-feeding in the stomach, rats being prevented from eating via the mouth. Rats learned to press a lever to obtain food in the stomach through a nasopharyngeal catheter (Epstein & Teitelbaum, 1962; Snowdon, 1969; Snowdon & Epstein, 1970). Under these conditions they developed a self-feeding pattern and even responded to a dilution of the liquid diet. However, it was shown that saccharin applied to the tongue during self-intragastric feeding facilitates the learning and 'energizes' the response (McGinty *et al.*, 1965). It was also shown that the nasopharyngeal catheter provides a pharyngeal stimulation in the throat. When this stimulation was eliminated, rats did not learn the self-intragastric feeding (Holman, 1969). Thus, it seems that stimuli from a lever and pressing the lever cannot be substituted for oro-pharyngeal stimulation by the food and for the oral motor pattern.

In human subjects, self-intragastric feeding allows the subjects to acquire, in three sessions per day, the amount of calories needed to maintain their weight; but they respond badly to a caloric dilution or to liquid food pumped into their stomach in addition to their oral intake. When exclusively self-fed intragastrically, they said that they suffered a lack of oral satiety (Jordan, 1969). The shift from the initial palatability of the food to its unpalatability at the end of the meal has been studied by recordings of chewing–swallowing patterns. Numbers of masticatory movements and durations of chewing before swallowing per unit of food are indices of palatability (Bellisle & Le Magnen, 1980). From these indices, differential palatabilities of various foods may be assessed at the beginning of a meal initiated in a state of constant deprivation. Regardless of the initial level, these indices indicate a decrease in palatability in the last compared to the first quarter of the meal (Bellisle & Le Magnen, 1980). These objective indices of palatability can be correlated in each subject to his/her judgement of pleasantness or unpleasantness (likes and dislikes), according to a psychophysical scaling. The shift from pleasantness to unpleasantness of food sensory stimuli, caused by eating the food, is a daily experience of human beings. It has been extensively studied by psychophysical evaluations and designated 'negative alliesthesia' by Cabanac (1979).

Stomach

The role of a combined action of the mouth and the stomach, excluding the role of intestine, has been the subject of numerous experiments. That food accumulating into the stomach alone can block a sham-feeding or real feeding via the mouth has been confirmed by many studies. If the emptying of the stomach was prevented by a pyloric ligature, food put into the stomach inhibited both sham-feeding and real feeding (Kraly & Smith, 1978; Deutsch & Gonzalez, 1980; Deutsch *et al.*, 1980). In other experiments, food entering the stomach in rats or monkeys eating until satiation, is withdrawn by a gastric fistula or, instead, saline or nutritive solutions are added to the stomach contents through the cannula (Deutsch & Gonzalez, 1980). In rats, such manipulations led Deutsch & Gonzalez to conclude that the stomach alone is involved in the satiation process and that two sensing mechanisms may be involved at this level. The mechanical one, stomach distention, would play a part in determining the upper limit of the meal. Below this limit, the inhibition of the oral intake would be controlled by a sensor not of the volume of food in the stomach but of the caloric content. The volume required to stop the oral intake increases in proportion to a dilution of the stomach contents, e.g. by adding saline through a gastric cannula

(Deutsch & Gonzalez, 1980; Deutsch *et al.*, 1980). For the monkey, similar conclusions were drawn (McHugh *et al.*, 1975; McHugh & Moran, 1978; Wirth & McHugh, 1983). Food-deprived monkeys drank a liquid diet (0.5 kcal/ml) and reached satiety in bouts of drinking of 15 min duration. If the liquid food was then withdrawn from the stomach, they immediately resumed their drinking. In four successive bouts, re-started by the withdrawal of food, they drank twice the volume that was drunk in four successive spontaneous bouts. The amount of food leaving the stomach to go to the intestine did not account for these differences and it was concluded that, in the monkey, stomach distention is the only determining factor involved. Previous experiments of stomach loads in the monkey showed a perfect compensation by a reduction in oral intake. Results from other investigators, however, also in the monkey, suggest that the intestine, not the stomach, plays the major role in satiation (Gibbs *et al.*, 1981).

Intestine

By infusing various nutritive solutions into the duodenum during sham-feeding or real feeding, many investigators thought to demonstrate the exclusive role of the intestine in the satiation process. However, there was convincing evidence for a combined action of food in the stomach and of stomach emptying toward the intestine during the meal. During real feeding of the rat, a liquid food was infused through a fistula into the duodenum at a rate varying from 0.06 to 0.44 kcal per min. These infusions reduced the oral meal size and the 60 min intake in a dose-dependent manner. An infusion of 2 to 3 kcal stopped the sham-meal at a threshold rate of 0.11 kcal per min. When the transit from the mouth to the intestine was re-established by closing the fistula, the rate of stomach emptying during the meal was 0.3 kcal per min and the total emptying at the end of the meal was 3.8 kcal. It was also demonstrated that the rate of stomach emptying of liquid food of various densities is a constant number of kcal per min; this again suggests an as yet unidentified sensing of the number of calories by the stomach (Kalogeris *et al.*, 1983; Reidelberger *et al.*, 1983).

The exact nature of the participation of the intestine in the satiating process has been questioned. A theory fashionable some time ago was the possible involvement of the intestinal release of cholecystokinin (CCK), one of the numerous gastrointestinal hormones. The proposal was based essentially upon the fact that injections of CCK at pharmacological doses, administered intravenously or intraventricularly, inhibited feeding; but such an effect does not prove a physiological action. The injection of many agents and drugs by various routes either inhibits or facilitates feeding and there is no evidence for their physiological involvement.

In rats, administration of the terminal octapeptide fragment of CCK at physiological doses at the end of a meal stimulates intake at night and inhibits it slightly during the day (Smith & Gibbs, 1975; Kraly, 1980). In elegant studies by Koopmans (1981), 30 cm of the small intestines of two rats were crossed over. Food eaten by the first animal leaves its stomach and enters the intestine of the second rat, then comes back to the first one, and vice versa. The first rat, presented with food, eats a meal of normal size. Because CCK could not be released into its intestine, the actions of the intestine and of CCK in determining satiation are excluded. Oral intake is also unchanged in the other rat if offered food sometime afterwards. Thus CCK exerts no satiating effect. Finally, CCK at pharmacological doses clearly stops peristalsis and the observed inhibition of feeding is the result of a trivial conditioned taste aversion (Deutsch & Hardy, 1977; Deutsch *et al.*, 1978; Bernstein & Goehler, 1983; Swerdlow *et al.*, 1983). Also, in a sham-feeding associated with CCK administration, the normal meal size to post-meal satiety correlation was not reproduced (Kraly, 1981).

The role of intestinal emptying towards the portal vein and the liver by the intestinal absorption has been very little investigated. Interestingly, however, it has been shown that a blockade of this absorption produces a long-lasting satiety. Adding mannitol (a non-absorbable agent) to the food in various concentrations results in a long-lasting accumulation of the ingested food in the upper intestine and the entry of water from the systemic compartment. The size of the meal is not affected by this situation, whatever the concentration of mannitol; but the time elapsed until the onset of a new meal is augmented in proportion to this concentration (Bernstein & Vitiello, 1978). This important result seems to indicate that, unlike gastric emptying, intestinal emptying during the meal is not involved in determining meal size by contributing to satiation. But, it is suggested that the intestinal distention and fulfillment maintain an active and long-lasting peripheral inhibition of oral feeding. The prevention of gastric emptying by a permanent pyloric ligature would presumably cause a same effect. This 'peripheral satiety' prevents the onset of the normal post-absorptive process of the meal (Bernstein & Vitiello, 1978).

Finally, two major points regarding the gastrointestinal satiation must be noted. First, only volume and number of calories are implicated. Despite some discrepancies it seems that, during the meal, pre-loads and loads of fats, proteins or carbohydrates, which may affect differentially the post-prandial satiety, act in satiation only through their caloric value

(Booth, 1972*b*; Maggio & Koopmans, 1982; Geliebter *et al.*, 1983). Secondly, the volume of food or the amount of calories needed at the gastrointestinal level to balance the oro-pharyngeal stimulation, i.e the palatability of the food, must be of the same degree as the hunger-dependent palatability. This explains of course the relationship between food palatability and meal size, when foods of different palatability are compared. A glucose infusion into the duodenum inhibits the licking rate of rats presented with a low palatability 2% glucose solution. The same infusion did not alter the rate of licking milk (Campbell & Davis, 1974; Davis *et al.*, 1976).

Pre- and post-absorptive events in satiation and satiety

Eating a meal results in the filling of a gastrointestinal store. All the above data point to the fact that this filling during the meal achieves satiety before intestinal absorption of the food because of a peripheral negative feedback counteracting the initial positive feedback of the hunger-dependent palatability. It will be seen later that this peripheral satiation mechanism is neurally mediated and dependent on afferent pathways to the brain (Chapters 8 and 9).

When and how is the systemic signal (or hunger signal), which has started the meal, turned off? The replacement of peripheral satiation by a persistent satiety necessarily requires the abolition of the stimulus which has initiated the meal. In as much as this initiation is due to the immediate beginning of a failure of food supply to the liver, satiety is therefore logically due to the reversal of this situation, i.e. to the intestinal absorption during or after the end of a meal. In rats, the first signs of the prandial post-absorptive phase (rise of blood glucose and of plasma insulin concentration) occur 8 to 10 min after the beginning of a meal. This 10 min delay is critical and has been noted by various authors. A pre-load in the stomach inhibits oral intake after only 10 min delay. In the *ad libitum* condition, the rat's meal rarely exceeds 10 min duration. The post-absorptive rise in glycaemia and insulin may be the turning-off of the stimulation to eat, i.e. the systemic contribution to the onset of satiety. The inhibition of this post-absorptive insulin release by mannoheptulose inhibits the satiating effect of a pre-load of glucose solution (Booth & Jarman, 1975). In the pig, injecting insulin during the meal reduces meal size (Anika *et al.*, 1980).

However, are pre-absorptive systemic events involved in determining meal size? The finding of the cephalic phase of insulin release has provided evidence for a possible role of this conditioned endocrine response. Within the first one or two minutes of the beginning of oral intake in rats, an abrupt rise of plasma insulin concentration and subsequent drop of blood glucose were observed (Louis-Sylvestre, 1976). This phenomenon discovered in the 1970s in the dog (Fischer *et al.*, 1972) and the rat (Louis-Sylvestre, 1976), and later in humans (Bellisle *et al.*, 1983), has been designated the 'cephalic phase of insulin release'. It was shown by Louis-Sylvestre (1976) and Le Magnen (Louis-Sylvestre & Le Magnen, 1980*a*) that this pre-absorptive endogenous release of insulin by the pancreas is induced from the mouth as a reflex. A subdiaphragmatic vagotomy or the transplantation of pancreatic islets to the kidney after the destruction of pancreatic beta cells by streptozotocin abolishes the insulin release at the beginning of the meal. It is orally mediated. A saccharin solution, like a sugar solution, elicits the response. It is 'a conditioned reflex' in as much as it may be extinguished (Deutsch, 1974) or modified by changing the food palatability (Berridge *et al.*, 1981). The amount of released insulin is dependent on the palatability of the food, this palatability being judged by meal size. Rats presented with three versions of the same diet each made palatable in different ways ate, during one meal, 4.9, 3.1 and 1.5 g of the high, medium and low palatability food, respectively. The rise of insulin concentration recorded at the beginning of the meal was, respectively, 65, 46, 28 μU/ml (Louis-Sylvestre & Le Magnen, 1980*b*). This parallelism of the two correlations between, on the one hand, palatability and the pre-absorptive insulin and, on the other hand, palatability and meal size is not surprising. The induction of insulin release as a reflex at the beginning of the meal, and the resulting hypoglycaemia, add their effects to that of the systemic level which has initiated the meal. This acts to enhance the hunger-dependent palatability of the food at the start of the meal. It is reflected by a slight and transient acceleration of the eating rate during the first few minutes of meals in mice (Wiepkema *et al.*, 1966) and in rats. In man, it is reflected by the increased feeling of hunger at the beginning of a meal expressed in the popular adage in French: *L'appétit vient en mangeant*. Thus affected by the pre-absorptive release of insulin, the level of palatability also determines meal size (Louis-Sylvestre, 1983*b*).

However, further study will be needed to ascertain if there is a possible action of the pre-absorptive events on the size of the meal. In the present state of knowledge, it is possible to assume tentatively that factors which induce insulin release as a reflex during the pre-absorptive phase tend to augment meal size, whereas hyperglycaemia and insulin release at the onset of the post-absorptive phase contribute to the satiating process.

5 Body energy balance

In animals and humans, it may be commonly observed that the total energy utilization, i.e. energy expenditure plus energy retention for growth, gestation or lactation, is balanced by an equal amount of energy intake. The constant species-specific and genetically programmed amount of the body fat mass is the result of this equilibrium between the input and output of body energy. Before examining in the next chapter how this constant level of body fats is preserved by a liporegulatory mechanism, we will look at whether and how it is preserved by an adjustment to each other of energy inflows and outflows. This can be examined when both energy expenditures and the caloric content of the diet are in a steady-state condition or in dynamic conditions of change of expenditure and of free food intake.

The interpretation of experimental data on the topic have often been, and are still, obscured by speculation. Theoretically, regulation of body energy balance could be achieved through one of three different models. One concept of output regulation is that the equilibrium between energy input and output might be achieved by an increase or decrease in thermogenesis, respectively added to or subtracted from obligatory energy expenditures which would compensate for random and uncontrolled variations in food intake. Another theory is that food intake might be dependent on changes in thermogenesis through a sensing either of heat outflow or of O_2 consumption. In fact the experimental data reviewed in this chapter have enabled these two models to be excluded and demonstrate that food intake is ultimately controlled by the turnover of the metabolic substrates of thermogenesis.

The steady-state condition

The dual periodicity of food intake set against the fluctuating but continuous energy output leads necessarily to short-term or middle-term imbalances. During a meal, of course, the rate of energy intake is considerably higher than energy expenditure. In rats and humans energy intake in a 1 h daily meal may represent energy expenditure over 24 h. In some animals species, a meal (for example a boa swallowing a small goat) can represent three weeks of energy expenditure. The high intake at night and low intake during the daytime in nocturnal species also leads to middle-term imbalances. The balance between input and output and their respective mean rates over time can only be realized through long-term adjustment.

Simultaneous, long-term recordings of O_2 consumption and free food intake in adult rats have verified these assumptions (Le Magnen & Devos, 1970, 1982, 1984; Le Magnen *et al.*, 1973). The computation of the meal-to-meal energy balance (caloric intake in a meal minus energy expenditure from the start of one meal until the next) shows excesses of intake at night and deficits during the daytime compared to the meal-to-meal energy output. During the night and particularly during the first 6 h, meals are initiated sooner than predicted on the basis of the utilization of the ingested food to cover only the recorded metabolic rate. During the daytime, meals are initiated later than predicted if the food ingested only supplies meal-to-meal expenditure. However, no positive correlation exists between caloric intake during meals and the post-prandial metabolic rate (ml O_2/min). Therefore, the bigger and smaller meals are not regulated during the two periods by increases and decreases in thermogenesis. Rather they are regulated by longer or shorter post-meal intervals, as assessed by the post-prandial correlation (see Chapter 2). But as substantiated by the difference in this correlation between the two periods, the total nocturnal intake is in excess of, and in daytime is lower than, the expenditure. Interestingly, the higher the excess at night, the lower is the deficit during the subsequent daytime. A correlation exists between nocturnal excesses and the total duration of no eating on subsequent days. This is, in the 24 h periodicity of feeding, the equivalent of the post-prandial correlation. The effect of this dark–light compensation is that the 24 h balance is close to an equilibrium between intake and expenditure. The remaining positive energy balance is positively correlated to the slight daily weight gain. As mentioned earlier, no cyclic fluctuation is observed in adult male rats beyond 24 h (Fig. 5.1).

Thus, in male rats and in the experimental condition of restrained activity and neutral ambient temperature, feeding mechanisms operate to adjust the cumulative discontinuous intake to cumulative expenditure, i.e. to the mean rate of output of energy within 24 h. Nothing indicates that the energy balance is maintained by adjusting expenditure to intake.

In man, the day-to-day relation between free food intake and daily metabolic rates has been followed in a series of subjects. No relation exists between intake and expenditure on the same day. Depending on the subject, total intake matches total expenditure only over days, weeks or more (Durnin, 1957, 1961). In inter-individual comparisons, it was observed that one subject can consume daily twice as many calories as

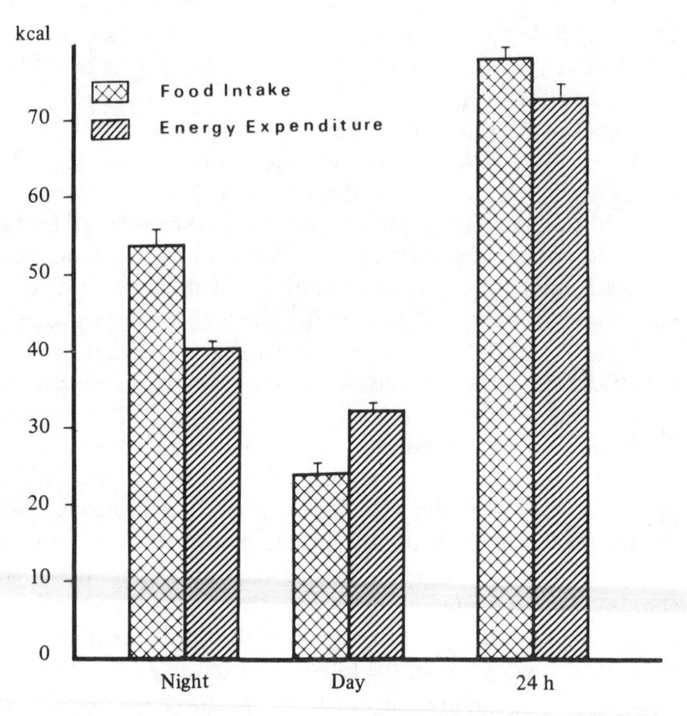

Fig. 5.1. Energy intake and expenditure during night, daytime and over 24 h. A positive energy balance at night is compensated for by a negative energy balance during the daytime.

another subject of the same age, weight and height (Widdowson, 1962). But it is also known that the metabolic rate of one can be twice as high as that of the other. Each human subject staying in a calorimetric chamber for 24 h was shown to intake food freely according to his/her 24 h expenditure. This was the case for adults with normal weight but not in obese subjects (Irsigler *et al.*, 1979).

These experimental data are inconsistent with thermostatic or energostatic theories which claim that the rate of energy flow *per se* would control food intake (Booth, 1972*b*). A sensor of the cumulative energy output is unconceivable. A sensor of the *rate* of output might be skin temperature, but a correlation between meal-associated changes in skin temperature and temperature in the brain has been refuted (Grossman & Rechtschaffen, 1967), as have changes of skin temperature as triggering factors of meal. Another sensor could be the O_2 consumption rate. But no relation has yet been found between carotid sinus functions and food intake. A change of metabolic rate just preceding a meal and its detection somewhere in the body are highly romantic notions.

Changing energy output

The two conditions of obligatory and non-obligatory changes in energy expenditure and energy retention must be distinguished. (Forced and 'voluntary'changes in food intake will be distinguished in the next section.) In various situations, food being freely available, augmentation and reduction in energy expenditure as heat or energy retention for lean-tissue synthesis are obligatory changes. In the cold and during exercise, while normal body temperature is maintained, any increase in thermogenesis is obligatory. The increase can only be balanced by either weight loss through the use of body fat reserves or an enhanced food intake. This is also true during growth, gestation, lactation and in changes in lean body mass in adulthood. Conversely, changes of metabolic rate can also be obligatory reductions. They are realized in hot weather, after parturition, etc. In these cases only weight gain or reduction of food intake can achieve the body energy balance of lean tissue. Thus, the above series of possible imbalances cannot be used to test whether energy balance is achieved by a modulation of output or of input. Since the choice is only between weight gain/loss (i.e. increase or decrease of body fats) and voluntary changes in food intake, these conditions will be examined in Chapter 6, which is concerned with the regulation of the body fat mass.

Increases or decreases in random or general activity represent a non-obligatory change in expenditure. It was shown earlier that such changes in spontaneous activity may be involved in the restoration of body energy balance after exercise or food deprivation. But these changes in activity are not effectors of short-term adjustments of expenditure to intake. As mentioned above, big meals are not followed by a higher metabolic rate, which would suggest hyperactivity. Rather, the results point to hypoactivity as being the behavioural symptom of satiety.

Changing food intake

Forced and restricted feeding result in conditions which are parallel to the obligatory changes in thermogenesis. Forced overeating and undereating can only be balanced by weight gain/loss or by enhanced or reduced regulatory changes in thermogenesis. So we can look for liporegulatory mechanisms in these situations. Only cases of excessive or deficient voluntary oral intake, and their eventual balance through a change of non-obligatory expenditure, will be examined.

In rats fed the same, familiar food *ad libitum*, this condition of overeating or undereating is never realized. That feeding has to be overstimulated or restricted to yield any evidence for regulation by thermogenesis, in fact suggests that the normal regulation of energy balance is achieved by the control of feeding.

Caloric adjustment

Changing the constant familiar food allows us to test the capacity of animals and human subjects for adjusting their intake to energy expenditure or to change the heat production to match an excessive or deficient intake. The adjustment to a change in the caloric density of the diet by an increase or decrease of the ponderal intake is well established in rats, rabbits, dogs and monkeys (Gasnier *et al.*, 1932; Hansen *et al.*, 1971). Indeed it is so well-established that it is not, as are many other phenomena, periodically re-examined. Adding an inert material, cellulose or kaolin, to the familiar stock-diet, readily induces a ponderal augmentation of the free intake which compensates exactly for the reduced caloric content. This adjustment is achieved within two or three days (Janowitz & Grossman, 1949; Janowitz & Hollander, 1955). If 50% of the diet is inert material, adjustment is impaired and spillage occurs presumably as an effect of the change in palatability. The increase in *ad libitum* intake is performed initially mainly by a shortening of meal-to-meal intervals, and therefore by an increase in meal frequency, and also progressively by an increase in meal size (Le Magnen, 1969). Concentrating the food again might produce caloric overconsumption leading to obesity or might be corrected by induced thermogenesis. In fact, it is counteracted by a reduction of ponderal intake as rapid as the augmentation was in the case of caloric dilution. This adjustment also occurs in rats and dogs fed a single daily meal, but it is much slower and imperfect (Janowitz & Hollander, 1955; Jacobs & Sharma, 1969). On a feeding schedule of several meals per day at fixed hours, the adjustment partly fails in rats (Le Magnen, unpublished). This again gives evidence for distinct mechanisms controlling meal frequency and meal size and for an efficient regulation through the former. In rats, adding inert materials to the diet reduces simultaneously the caloric density and the palatability of the food. This altered palatability can impair the rat's adjustment by an increased intake, particularly in meal eating.

In human subjects, many experiments have shown that this adjustment to caloric dilution of the food is slow and imperfect (Wooley *et al.*, 1972; Spiegel, 1973; Spiegel & Jordan, 1978). A typical experiment is as follows (Porikos *et al.*, 1982): the intake of six subjects during their normal freely available meals was controlled for 24 days. Then a 25% reduction of the caloric density of the various courses was achieved by substitution of a non-caloric sweetener (aspartame) to items of the menu containing sucrose. The volume intake of subjects did not change during the first three days. Later and until the 20th day, their intake slowly increased. However, they only compensated for 40% of the reduction in caloric density.

Overstimulation of intake

A change in the palatability of the food without a simultaneous change in its caloric density and nutritive properties immediately augments or reduces the 24 h intake of free-fed and meal-fed animals. The persistence of this palatability-induced overeating or undereating could lead either to weight gain or weight loss or to a regulation by increase or decrease of energy expenditures. None of these is the case; the regulation here again is through the feeding mechanism, i.e. by a 'reconditioning' of palatability. This is manifested by the disappearance of the excessive or deficient intake. On a diet made more palatable by non-caloric sweeteners or by an inert greasy material the normal intake is rapidly re-established. It is the same with a diet adulterated by addition of a bitter-tasting compound such as sucrose octoacetate. As already mentioned, the persistence of low intake with a quinine-adulterated diet (p. 41) has been shown to be the effect of the post-ingestive toxic effect of quinine and therefore of a conditioned taste aversion (Kratz *et al.*, 1978; Aravich & Sclafani, 1980). In this last case, weight loss is recorded.

When rats are presented with a permanent choice of high palatability foods, overeating is as transient, as it is with a single high-palatability food (Louis-Sylvestre & Le Magnen, 1984). However, daily changing of the high-palatability or low-palatability food, which prevents the post-ingestive modulation of palatability, produces hyperphagia. This condition designated *cafeteria regimen* has been the object of many experimental studies dealing with body energy regulation (e.g. Sclafani & Springer, 1976; Rolls & Rowe, 1977). The conclusions drawn by various authors are contradictory and the subject is highly controversial (Louis-Sylvestre & Le Magnen, 1984).

Two points must be noted initially. The overeating of rats placed in a cafeteria regimen is a 'meal overeating'. Rats maintain their nocturnal intake at a higher level than the diurnal one: but at night, they take huge meals, during which they take the various foods successively and alternately (Rogers & Blundell, 1980; J. Le Magnen & M. Devos, unpublished). Thus, the overeating is an effect of 'variety' and of the sensory-specific stimulation and satiation, already discussed (see Chapter 3). A second point is that rats offered a series of human foods (also designated 'supermarket foods') select a daily diet which differs nutritionally from their stock-diet. This selected diet is a high fat diet.

This cafeteria-induced overeating seems to be a deviation from regulation by adjustment of food intake to metabolic demand. Consequently, under these conditions rats either will become obese or will maintain their body weight. In the latter case, it should be suggested that extra heat production has compensated for the excessive energy intake. Rats become obese and do indeed gain more weight than controls (19% vs 14% in 10 days). However, some data tend to demonstrate that this weight gain, though not prevented, is at least slowed by a diet-induced thermogenesis involving the brown adipose tissue (BAT). Other results suggest the opposite; that is to say that the weight gain accounts entirely for the augmented caloric intake.

The first suggestion of an extra heat production came from calculations of the 'metabolic efficiency' of the food for weight gain. The computation of the ratio of total caloric intake to weight gain (in grams) used in veterinary studies of food efficiency for growth is misleading if used for our calculations. An enhanced ratio in cafeteria-fed rats does not suggest convincingly that part of the extra calories was dispersed as extra heat. A direct demonstration of this extra heat production and of the involvement of BAT was required. A series of elegant experiments seems to have provided this (Rothwell & Stock, 1979; Brooks *et al.*, 1981).

The body energy gain per gram of food intake was reduced by 39% in cafeteria-fed rats compared to controls. Intermittent and short-term recording of resting metabolism indicated a 25% increase in O_2 consumption. Noradrenaline injections increased the O_2 consumption of cafeteria-fed rats more than that of the controls, similar to the case in cold-acclimated rats. Also as in cold-acclimated rats, the weight of BAT was shown to be augmented in cafeteria-fed rats. This tissue was hypertrophied. The increase in BAT metabolism alone accounted for the whole body measurement of increased metabolic rate. The binding of guanosine diphosphate on the internal membrane of BAT mitochondria (which is an index of the thermogenic activity of the tissue) was three- or four-fold that of controls after 3 days of cafeteria regimen. After excision of intrascapular BAT, cafeteria-fed rats gained more weight than unoperated controls (Stephens, 1981).

Despite this and other convincing evidence for the activation of an extra heat production by the BAT in cafeteria overeating, similar to that induced by exposure to cold, a discrepancy exists regarding the measured body energy balance of such rats. In a particular experiment the body energy balance of rats fed for 15 days on a cafeteria regimen, and compared to controls fed by stock-diet, was established by measurement of the V_{O_2} consumption during the treatment and by a measure of lipids in the carcass of the two groups sacrificed before and at various steps in the treatment. V_{O_2} total consumption minus the energy retention during the 15 days confirmed a 40% increase of thermogenesis under the cafeteria regimen. The calculated cost of fat synthesis (30% of its caloric value) accounts for only 10% of this measured increase in thermogenesis. The rest was attributed to the extra heat production by the BAT (Rothwell & Stock, 1982). Identical determinations in the same strain of rats of the same age by two other research groups led to the opposite conclusion. In cafeteria-fed rats daily thermogenesis was increased by 24%. The exothermic cost of the measured deposed fats accounted for this thermogenesis (Armitage *et al.*, 1979, 1981). Barr & McCracken (1982; McCracken & Barr, 1982) computed that only 7 kJ of the heat production per day and per rat were in excess of that implicated in fat synthesis. These authors observed the increase in the weight of BAT, but they have shown that it is proportional to the weight gain of white adipose tissue (WAT) (Bestley *et al.*, 1982; Hervey & Tobin, 1982). Indeed a two-fold increase in BAT during cold exposure bears no relationship to the increase or decrease in WAT.

In man, direct or indirect calorimetry did not support a diet-induced thermogenesis in people of normal weight, nor its deficiency as a possible cause of obesity. The higher metabolic rate in obese subjects is due to their weight as well as to their higher extra heat production during exercise (Blaza & Garrow, 1983). In humans, as in animal models, the participation of a modulation of thermogenesis in body energy balance will have to be re-examined in a condition of forced overeating and food restriction.

6 The regulation of body weight or body fat mass

All vertebrates possess a large-capacity species-specific store of endogenous energy. This store, distributed in various depots throughout the body, is the white adipose tissue. As is well known, within adipocytes energy is stored as triglycerides, molecules able to produce the highest amount of metabolizable energy per unit weight (9 kcal/g). Synthesis of triglycerides (lipogenesis), and hence the filling of the store from plasma glucose and triglycerides, is insulin dependent. Their mobilization as free fatty acids (lipolysis), and hence the emptying of the store, are dependent on the autonomic nervous system through direct adrenergic innervation of the adipose tissue, catecholamine release from the adrenal medulla and neuronal control of the endocrine pancreas (particularly its glucagon release).

Short-term imbalances between continuous and fluctuating energy outflow and intermittent feeding are buffered (as described in the preceding chapter) by the prandial store and its utilization from meal to meal. Middle- and long-term imbalances are buffered by the endogenous fat store. Transient or persistent conditions of positive energy balance, energy intake being greater than expenditure, lead to fat synthesis; negative energy balance, energy intake being less than expenditure, leads to mobilization and utilization of fat as fuel for energy metabolism. In the first case, lipogenesis is the transfer of energy metabolites from the small-capacity gastrointestinal store to the large capacity adipose tissue store. Then the total rate of utilization of energy intake is the sum of the rates of energy metabolism of lean tissue and of lipogenesis. In lipolysis, on the other hand, the rate of metabolic utilization of food is the rate of energy metabolism of the lean body mass minus the endogenous caloric supply from fats.

Nevertheless, except for pathological conditions of obesity and leanness, a middle- or long-term maintenance of the body fat mass and of body weight is observed. This implies that, in addition to mechanisms which tend to achieve body energy balance by adjustment of intake to expenditure or vice versa, the body fat content is regulated. In other words, at some defined limit an increase in body fat mass is corrected by a subsequent depletion, while its occasional depletion is corrected by subsequent addition. Experimental data reviewed in this chapter fully confirm these suggestions and detail the mechanisms involved.

However, it is useful to recall that the regulation of body fat mass is not identical with the regulation of body weight. Body weight *per se* is not regulated of course. Astronauts, after a month of weightlessness, do not gain 500 kg to compensate for the total loss of their weight. Body fats are only a component of body weight. Other components, body water and lean body mass, are regulated by mechanisms independent of those involved in the regulation of adipose tissue mass. Therefore, fluctuations in body weight may be taken as fluctuations in body fats only when other elements of the total weight of the body, particularly the lean body mass, are maintained. When possible, it is better to use a post-mortem measurement of lipids in the carcass to assess changes in the fat stores.

The liporegulatory mechanism and its impact on *ad libitum* feeding

In the early 1970s, the simultaneous recording of 24 h food intake and respiratory exchanges in rats pointed to a liporegulatory mechanism in the 24 h body energy balance and its determining effect on the nycthemeral periodicity of feeding (Le Magnen & Devos, 1970; Le Magnen *et al.*, 1973). The following evidence showed that nocturnal positive energy balance was associated with fat synthesis and weight gain: high respiratory quotient, the increase of the lipid content of the carcass from the beginning to the end of the night (Panksepp, 1973), the high uptake of labelled acetate in adipose tissue (Kimura *et al.*, 1970) and the weight gain (Kakolewski *et al.*, 1971) (Fig. 6.1). At night the rate of fat synthesis added to the rate of energy metabolism accounted for the high ratio of meal intake to post-prandial interval, in other words for the acceleration of successive initiations of meals. During the daytime, the opposite was observed. The negative energy balance was associated with a low respiratory quotient indicating lipolysis. The weight gained during the preceding night was lost during the daytime. This contribution of an endogenous fuel to energy metabolism accounted for the long post-meal intervals and for their being relatively independent of meal size. The day-to-day correlation between nocturnal positive and daytime negative balance was substantiated by the finding of a similar correlation between nocturnal fat synthesis and subsequent daytime fat mobilization.

The use of radioactively labelled food demonstrated that the endogenous fuel utilized during the day came from the food eaten in excess the preceding night and was indeed responsible for the delay of meal onset (Le Magnen *et al.*, 1973). During the night, rats were offered a

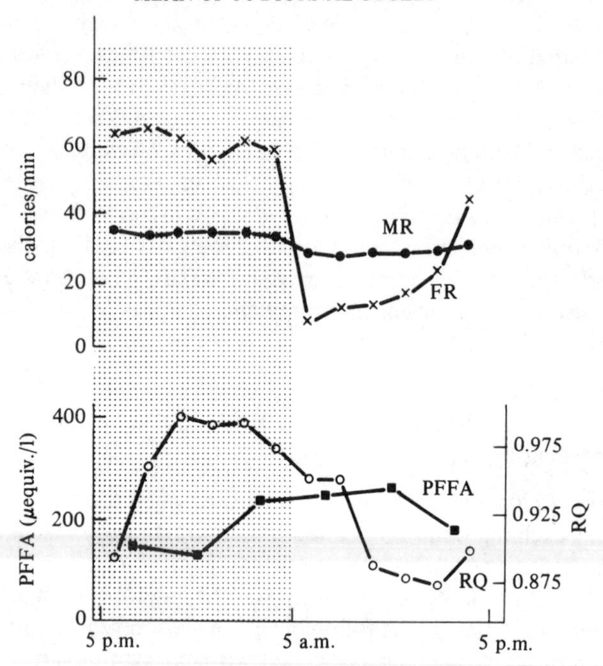

Fig. 6.1. Feeding rates (FR) which are higher at night and lower during the daytime than concomitant metabolic rates (MR) lead to fat synthesis at night and fat mobilization during the daytime, as assessed by high and low respiratory quotients (RQ) respectively. PFFA = plasma free fatty acid.

synthetic diet labelled by [^{14}C]glucose (1.8 μCi/100 diet units), and a similar, unlabelled diet during the subsequent daytime. The recording of $^{14}CO_2$ expiration showed that only 72% of radioactivity consumed during the night were expired during that time. The remainder was expired during the subsequent daytime, indicating the utilization of calories stored the preceding night in energy metabolism. This utilization, as assessed by an augmentation of $^{14}CO_2$ expiration, increased with each meal-to-meal interval and reached a maximum level at the onset of meals (Fig. 6.2). When other rats were fed the labelled food during the daytime, all the radioactivity was recovered during this period and the maximum $^{14}CO_2$ output was recorded just following, instead of just prior to, each meal.

The above results confirm that the nocturnal hyperphagia and diurnal hypophagia exhibited by free-fed rats are secondary effects of the filling and emptying, respectively, of the body fat reserve. There is other evidence supporting this finding: the neuroendocrine pattern

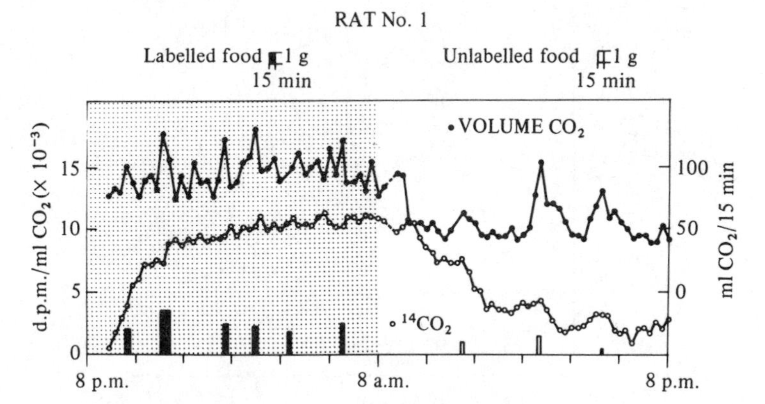

Fig. 6.2. $^{14}CO_2$ coming from oxidation of [^{14}C]glucose eaten by the rat during the preceding night is expired during the daytime and increasingly prior to spontaneous meals.

underlying fat synthesis and mobilization (described below) is attenuated and persists under fasting or food restriction (Apfelbaum *et al.*, 1972*b*). In man at the beginning of the day and in mice at the beginning of the night, the appearance of these neuroendocrine characteristics precedes the change in feeding pattern (Malherbe *et al.*, 1969; Petersen, 1978). A nocturnal infusion of adrenaline, which abolishes lipogenesis, immediately induces a daytime-like feeding pattern (Danguir & Nicolaïdis, 1979). And conversely, infusing insulin during daytime, which abolishes lipolysis, immediately induces a nocturnal-like feeding pattern (Larue-Achagiotis & Le Magnen, 1979). Thus, it is suggested that the liporegulatory mechanism is an autonomous control system, independent of the feeding mechanism, which indirectly affects feeding by increasing or decreasing the rate of glucose utilization.

What is the mechanism underlying this alternating primary lipogenesis and lipolysis which compensate for each other and lead to a middle- or long-term maintenance of body fat levels? The answer has been revealed in rats and in many studies in humans. The level of fat synthesis and therefore of the amount of glucose diverted towards this synthesis is dependent on endogenous insulin release by the pancreatic islets. As one might expect, both basal and glucose-stimulated insulin release were found to be higher during the lipogenic nocturnal phase in rats (Fig. 6.3) (Louis-Sylvestre, 1978*b*). Thus, for the same amount of the same food, the increase of plasma insulin concentration in the post-absorptive phase of each meal is higher at night than during the day. As a result of this high responsiveness of pancreatic beta cells to glucose at night, rats exhibit a nocturnal high glucose tolerance. After gastric or intra-

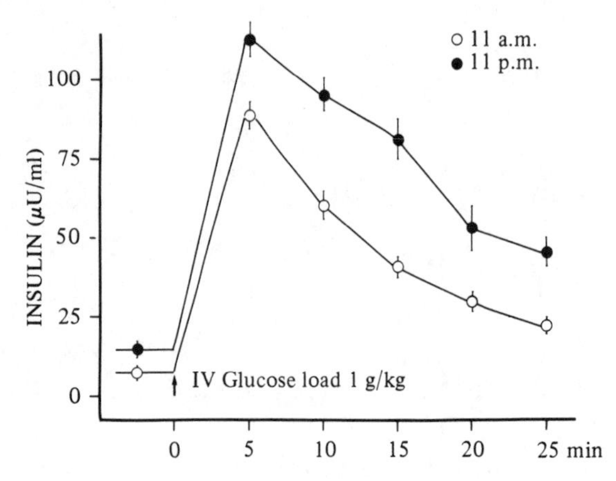

Fig. 6.3. Basal plasma insulin concentration and glucose-stimulated insulin release are higher at night than during the daytime in free-fed rats.

venous glucose loading, the rate of disappearance of glucose from the blood is faster at night, and a relative glucose intolerance is shown during the day (Le Magnen *et al.*, 1973). The return to normoglycaemia and further to hypoglycaemia after a glucose load, as after a meal, is accelerated at night. This, as was suggested above, accelerates the frequency of meal initiations by giving the triggering signal for initiation. Injecting insulin elicits a steeper fall in blood glucose level at night (Penicaud & Le Magnen, 1980*a*). This suggests that the enhanced endogenous insulin release is accompanied by an increase in the sensitivity of the target tissues (e.g. adipose tissue) to the action of insulin on glucose disposal.

This endocrine hormone pattern is neuronally mediated and (as will be seen in Chapter 10) governed by the central nervous system. Sympathetic fibres afferent to the beta and alpha cells of the pancreas have been shown to inhibit the secretion of insulin and facilitate that of glucagon. Activation of the vagal fibres to the islets facilitates insulin release. Secretion of the two glucoregulatory hormones in response to hyperglycaemia and hypoglycaemia is thus dependent on the balance between sympathetic and parasympathetic innervation of the islets. *In vitro*, it has also been shown to be dependent on the balance between the sympathetic and parasympathetic neurotransmitters noradrenaline and acetylcholine (Campfield & Smith, 1983). Subdiaphragmatic vagotomy makes glucose-stimulated insulin release during the night identical with that observed during the daytime (Campfield *et al.*, 1983), thus confirming that the nocturnal metabolic pattern is stimulated vagally. During the daytime, hypoinsulinism and glucose intolerance indicate a prevailing sympathetic activity underlying lipolysis. Thus the alternating metabolic and feeding nycthemeral pattern is also the alternation between a prevailing parasympathetic activity at night and prevailing sympathetic activity during the day.

At much the same time as these findings were made in rats, similar results were obtained with humans. All the nocturnal metabolic characteristics of the rat were found to be present during the daytime in man: the high glucose and food-stimulated insulin release, high glucose tolerance followed by nocturnal glucose intolerance, high tissue sensitivity to exogenous insulin (Malherbe *et al.*, 1969; Sensi *et al.*, 1973; Aparicio *et al.*, 1974; Gibson *et al.*, 1975; Sensi & Capani, 1976). The nycthemeral cycle of responses to the glucose tolerance test was discovered in 1967 (Specchia *et al.*, 1967) and was ignored by specialists in feeding.

So in a programmed alternation, 12 h fat synthesis and stimulated hyperphagia resulting in a loading of the body fat store is followed by emptying of this store during the next 12 h (of the day in rats and of the night in man). This last period is that of rest and sleep. This alternation during the 24 h periodicity is reminiscent of cycles of feeding and anorexia of longer duration in various species (Mrosovsky & Sherry, 1980). In the 2 month cycle of deermice, neuroendocrine characteristics of weight gain and weight loss phases identical with that of the 24 h cycle in rats and humans have been demonstrated (Melnyk *et al.*, 1983). The liporegulatory mechanisms in action in a programmed cycle are also concerned in buffering transient imbalances between energy intake and expenditure, e.g. resulting from cold exposure or exercise. It is also the case in spontaneously occurring hyperphagia. In all these conditions the depletion of fat stores is compensated for by subsequent repletion. Thus the observed long-term invariance in body fat levels is maintained.

Reversibility of conditions of spontaneous or forced overweight and underweight

The role of this autonomous liporegulatory mechanism in being complementary to the feeding mechanism proper has been further demonstrated during forced over- and under-feeding and conditions of forced overweight and underweight.

In cafeteria-fed rats the return to stock-diet is followed by an abrupt weight loss associated with hypophagia or aphagia. Rats lose a large part of the weight gained on the cafeteria regimen during the first few days of re-feeding on their standard diet (Sclafani & Springer, 1976;

Rothwell & Stock, 1979). When the initial body weight has been approximately re-established, rats resume their normal feeding pattern. This reversibility of cafeteria-induced obesity is upset when the varied diet is maintained for a long time (Rolls & Rowe, 1977; Mandenoff *et al.*, 1982). The resultant persistent and uncorrected obesity could be because after some weeks the cafeteria-induced weight gain is due not only to adipocyte hypertrophy but also to a developing and irreversible hyperplasia of the adipose tissue.

Experimental overeating and induced overweight have provided fully convincing evidence of the liporegulatory control. A forced overeating and experimental obesity were observed in rats and studied by three different means: gavage or stomach tube-feeding (Cohn & Joseph, 1962), electrical stimulation of lateral hypothalamus (Steinbaum & Miller, 1965), injections or infusions of insulin (Hoebel & Teitelbaum, 1966). As a result of chronic tube-feeding of twice the normal intake for some weeks rats can increase their adult body weight two-fold. As in the cafeteria-induced overeating, it was claimed that the weight gain with tube-feeding was reduced by diet-induced thermogenesis. Calorimetric measurements in rats tube-fed twice their normal intake excluded this interpretation. The observed augmentation of the metabolic rate is accounted for by the exothermic cost of fat synthesis and by the body size (Armitage *et al.*, 1979; McCracken & Barr, 1982). Electrically stimulating the lateral hypothalamus (see Chapter 8) produces hyperphagia. Two sessions per day double the normal oral intake. Rats are totally aphagic between sessions. After some days, rats are obese. During the sessions of stimulation, food being available, rats are highly hyperinsulinic (Steinbaum & Miller, 1965; Steffens, 1975). As mentioned earlier, a chronic treatment by protamine-zinc-insulin (PZI) or a continuous infusion of regular insulin through an intravenous catheter produces dramatic overeating and rapid weight gain until obesity (Hoebel & Teitelbaum, 1966; May & Beaton, 1968; Larue-Achagiotis & Le Magnen, 1979). The efficiency of a dose of injected or infused insulin in promoting the weight gain decreases in proportion to that weight gain. A plateau of induced obesity is reached. Under continuous insulin infusion hyperphagia is achieved by more frequent meals and most of all by the total disappearance of the daytime hypophagia. Under these conditions rats consume the same amount throughout the day as they would normally eat only at night (Larue-Achagiotis & Le Magnen, 1985*a*) (Fig. 6.4).

In the three above cases of forced overeating and induced overweight, the discontinuation of the treatment is followed by weight loss and a return to the initial and regulated body weight. During this weight loss,

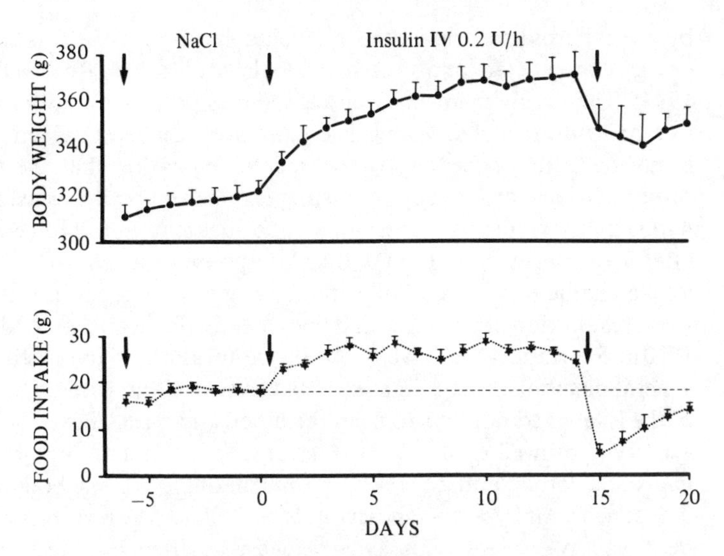

Fig. 6.4. Rats continuously infused by regular insulin via an intravenous (IV) catheter are hyperphagic and gain weight. The insulin-induced obesity is reversible. After 14 days, the cessation of treatment is followed by hypophagia and weight loss until the initial body weight has been re-established.

rats are hypophagic or totally aphagic. Their normal intake is resumed when the original body weight (before forced feeding) is re-established. During the weight loss after PZI treatment and insulin infusion, the residual food intake occurs at night and the meal pattern resembles that of normal daytime feeding with small and widely spaced meals (Le Magnen, 1976). During the weight loss there is a high plasma free fatty acid level, indicating lipolysis. Hypoinsulinism and glucose intolerance also indicate fat mobilization and metabolic utilization (Geary *et al.*, 1982; Carpenter & Grossman, 1983*c*).

Thus, over some weeks, forced weight gain followed by weight loss, and hyperphagia followed by hypophagia, mimic exactly the dark–light physiological pattern of rats. Thus, an experimental as well as physiological overrepletion of body fats can be corrected by a subsequent weight loss through the liporegulatory mechanisms.

The response to forced weight loss is a mirror image of the response to forced weight gain. As occurs tragically in some human populations, weight loss can be similarly realized experimentally in animal models by fasting or food restriction. It is equivalent to the overweight induced by tube-feeding. How free feeding is reduced under chronic treatment by a weight-loss inducing agent such as a catecholamine is unknown. During fasting, a reduction of the metabolic rate, presumably induced by hypothyroidism, slows the unavoidable progressive loss of body energy content (fats and proteins) (Boyle *et al.*, 1978; Rothwell *et al.*, 1982). The supply from endogenous reserves produces an abnormality in carbohydrate metabolism called 'starvation diabetes' which persists somewhat after re-feeding. It was mentioned earlier that the feeding response to one or two days of fasting is poor and that, at least in rats, hypoactivity seems to contribute to the rehabilitation. On re-feeding after a prolonged starvation and as a response to a substantial weight loss, a characteristic metabolic and feeding pattern leads in rats to the rapid restoration of body fats and body weight (Penicaud & Le Magnen, 1980*b*; Penicaud *et al.*, 1983). At the restoration of food access, the general metabolic rate is immediately elevated along with an elevation of the level of triiodothyronine in the blood and signs of catecholamine activity (Rothwell *et al.*, 1982). A strong lipogenesis and rapid weight gain are recorded until the re-establishment of the pre-fast body weight. The concurrent hyperphagia accounts entirely for the restoration of the body energy content. This hyperphagia is realized by a meal pattern comparable to that induced by insulin infusion in rats of normal weight. Under these conditions, rats are hyperinsulinic and display a high glucose tolerance, as do normal rats during the nocturnal period (Penicaud *et al.*, 1983). A high fat diet even of the same caloric density as the standard carbohydrate-rich diet, favours rehabilitation, presumably as an effect of the persistent disruption of carbohydrate metabolism (L. Penicaud and J. Le Magnen, unpublished). As mentioned earlier, rats offered a choice between three macronutrients augment their caloric intake by increasing only their fat intake (Andik *et al.*, 1951).

In humans, forced feeding has yielded contradictory results. In a classical experiment, volunteers ate 7 to 10000 kcal per day for some weeks. Some of them gained weight, as predicted on the basis of an unaltered metabolic rate; others did not. The former group lost weight at the cessation of the forced overeating and were hypophagic during this weight loss until the recovery of their previous body weight (Sims *et al.*, 1973). In another experiment, for 15 days subjects ate a daily excess of 1500 kcal over their normal consumption. Calorimetric measurements indicate an 11% elevation of the metabolic rate at the end of the period of overeating. The weight gain accounted for only half of the overeating; but only half of the remainder was accounted for by the elevation in O_2 consumption (Apfelbaum *et al.*, 1972). In another study, subjects were fed at three levels: high, middle and low. Recordings of metabolic rate by direct calorimetry showed an elevation of high intake by only 10%, and a reduction by only 6% on the low intake, compared to the middle one. These levels of change are compatible with the effect of intake on the heat increment of food which varies in proportion to the amount eaten (Dauncey, 1980). This heat increment of food must not be confused with diet-induced thermogenesis. However, there are indications that this extra heat production which accompanies the meal could be modulated in a regulatory way. It was found to be higher in anorexics and in people of normal weight than in obese people. This difference has been suggested to be a cause of obesity (Green *et al.*, 1975; Bessard *et al.*, 1982).

The origin of spontaneous obesity in some strains of rats and mice and in humans is beyond the scope of this book. It is possible to assume that obesity results from an impairment of the feeding or of the liporegulatory mechanisms or of their role in achieving body energy balance.

7 Developmental aspects

Ontogeny of feeding is a special topic which should be reviewed and discussed elsewhere. Here, it will be touched on briefly in order to provide an insight into the progressive emergence of phenomena and mechanisms (detailed in preceding chapters) during growth from birth to adulthood. This brief analysis will focus, as have previous chapters, on mammals, particularly the rat. Some data on human newborns will be also mentioned, the reader being referred to original works and relevant reviews.

Rat pups find, and attach to, a nipple of their mother and begin suckling in the first hours after birth. This initial attachment occurs without the assistance of the mother, as demonstrated by experiments with an anaesthetized mother. As the eyes of the young are closed, the possible roles of tactile, thermal and olfactory cues in the attachment have been investigated (Blass *et al.*, 1977). Shaving the fur around the nipples or cooling or heating the skin are without effect. Olfactory cues, however, are involved (Teicher & Blass, 1977). Attachment can be prevented by washing the nipples after parturition but it can be reinstated by coating the nipples with a distillate or an alcohol extract of the washing water. It can also be re-established by the amniotic fluid of the mother, which, by spilling onto the fur of the mother at parturition, seems to be a physiological stimulus (Teicher & Blass, 1977; Blass & Teicher, 1980). Saliva of the mother and not of a virgin female fed by the same diet also stimulated attachment when painted on the nipple after a previous washing. The saliva of the pups is also effective. However, 80% of pups survive after olfactory bulbectomy performed three days after birth though these anosmic pups do not have the same degree of capacity to attach. Olfactory cues can therefore be partly replaced by other signals (Teicher *et al.*, 1978).

Experience and learning seem to be involved in the appearance and evolution of attachment behaviour during the suckling period (Stoloff *et al.*, 1980). This behaviour would not, therefore, be fully developed at birth. Rats prevented from attaching during the first days and only permitted to suck intermittently are later persistently impaired in their attachment. However, attachment and even suckling are dispensible. Pups artificially fed through a gastric cannula, eat and grow normally after weaning. Furthermore, pups deprived of suckling are able to survive by eating sucrose, powder or mash spread on the floor of their cage (Hall *et al.*, 1975). Once initiated on the first day after birth, attachment to the mother's nipple is not driven in a constant reflex pattern. From birth to weaning at 20–21 days, an evolution has been observed towards an appetitive behaviour. Until 11–14 days of age, pups detached from the nipple have the same latency to re-attach whether or not they have been previously deprived of suckling for some time. Later, undeprived pups display a longer latency to re-attach than deprived ones. During the first two weeks they remain attached to the same nipple, at which they suck almost continuously, without any relation to milk ejection or availability. After 14 days of age, they detach spontaneously to change to another nipple, apparently in relation to milk availability.

One can observe a parallel evolution of the sucking pattern from the first to the last weeks of suckling, which demonstrates a progressive emergence of post-ingestive nutritional control. The recording of the sucking pattern by interposing a flexible cannula between the nipple and the mouth of the pup gave the following results: until 11–14 days, pups suck in intermittent bursts. This pattern is not affected by the periodicity of spontaneous or hormonally provoked milk ejection by the mother. In the litter as a whole, the sucking of pups is randomly distributed with respect to milk ejection (Wakerley & Drewett, 1975). Thus sucking does not influence milk ejection and vice versa. During this early period, a pup shifted successively to several foster mothers (each of them maximally stimulated by oxytocin treatment), takes an enormous volume of milk until its stomach is extremely distended. Pups deprived of suckling for 24 h will take milk amounting to 10% of their body weight from a non-stimulated mother (Cramer & Blass, 1983). In so doing, they demonstrate a response to deprivation which is not manifested in their latency of attachment. During this initial period, a load in the stomach, like spontaneous gastric filling, reduces further suckling; but an identical inhibition is obtained after a load of a glucose solution or of water (Hall, 1975; Houpt & Houpt, 1975). Later, when solid food begins to be found in the stomach, indicating a pre-weaning period, post-ingestive regulation of intake becomes conspicuous. Deprived pups only take milk equivalent to 5% of their body weight after 24 h deprivation. They continue to respond immoderately to an unlimited availability of milk but show a nutritional control of their intake in a condition of limited milk availability. At this time also, pups begin to give differential responses to nutritive and non-nutritive loads.

However, during the whole suckling period and for some weeks after weaning, rat pups do not respond to insulin (Lytle *et al.*, 1971) and to

2-deoxy-D-glucose (Houpt & Enstein, 1973; Gli-Ad *et al.*, 1975). At least in the early phase of development of the feeding system, this argues that the induction of feeding by insulin, and most of all by the cellular glucopenia produced by 2-deoxy-D-glucose, does not reproduce the physiological regulatory control of food intake which, as mentioned above, appears earlier. It should be noted that the cephalic phase of insulin release is present in rats as early as the first week of life (Bernstein & Woods, 1980). It will be seen below (Chapter 8) that lateral hypothalamic lesion induces aphagia in rat pups as it does in adults (Almli, 1978). However, other facts indicate a progressive maturation of the brain target of blood-borne systemic signals. This maturation is not fully completed at the time of weaning (Teitelbaum *et al.*, 1969; Cheng *et al.*, 1971).

The dark–light cycle of sucking is apparent early after birth. For 16–19 days, the mother rat seems to impose lactation on the pups during the light period. This is observed even with blind pups (Levin & Stern, 1975). Later, the adult-like nocturnal pattern appears with, initially, a very low night to daytime ratio (Bernstein, 1976*b*). The normal adult ratio is reached after 6 weeks during the post-weaning period. Until 6 to 7 weeks of age, during this post-weaning period, there is no correlation between meal size and post-meal intervals (post-prandial correlation; see Chapter 2) (De Castro & Balagura, 1976).

Conflicting results have been reported regarding the existence of a relation between meal size and pre-meal interval (pre-prandial correlation) before the appearance of the adult-type post-prandial correlation. This dependency of meal size on time elapsed since the preceding meal was found when rats were fed on liquid diets (De Castro & Balagura, 1976) and not on a solid diet (Bernstein, 1976*b*). During this pre-weaning period of rapid growth, the daily meal number is high and meals are small. Meal sizes later increase progressively, whereas meal number is reduced. This progressive development of meal eating seems to parallel the maturation of the lateral hypothalamic system. Lesion of this brain site prevents the meal patterning in young rats and abolishes it in adults (De Castro & Balagura, 1976).

Studies on the development of the differential palatability of foods in young rats and of early learning of food preferences and aversions have provided fascinating results.

Injection of an odorant into the amniotic fluid of a pregnant female followed by a LiCl injection, leads to an induced aversion to the compound being exhibited by offspring after weaning (Smotherman, 1982; Stickrod *et al.*, 1982). As assessed by the sucking pattern or by facial expressions, rats revealed unlearned palatability for sweet taste in the first days of life; and some days later, during the suckling period, the low palatability or a true aversion to a bitter-tasting compound. This aversion also seems unlearned but matures later.

The first demonstration of a transmission of food preferences from the lactating mother to rat pups was provided as early as 1968 (Le Magnen & Tallon, 1968*b*). Lactating mother rats were injected with an odorant daily for 10–30 days of lactation. After weaning, the young had to choose between a synthetic solid diet flavoured by the odorant and the same diet unflavoured. The young exhibited a highly significant preference for the form of the diet flavoured by the compound received by their mother during lactation. Later, Galef & Henderson (1972) and Galef & Sherry (1973) demonstrated, in a series of elegant experiments, the transmission, through the milk, of preferences based on olfactory cues associated with the food eaten by the mother. After weaning, rats prefer the diet flavoured like the diet eaten by their mother during lactation. It was demonstrated that this preference is transmitted by the milk as follows: 0.5 ml of milk were drawn from a lactating mother fed by a diet differing from that eaten by the mother of pup under investigation. This sample of milk was drunk by the pup and the intake associated with LiCl injection. After weaning, the rat exhibited an accentuated preference for the mother's diet and rejection of the diet of the lactating female, the milk of which was associated with LiCl-induced illness. However, two different factors were found to influence this initial choice. In addition to the associative learning which induces the preference for the mother's diet, the parental transmission was investigated. After weaning, young tend to eat the food preferentially in sites at which adults eat or have eaten. As in humans, young rats learn to like what their parents like or choose for them and this is presumably the case in many other animal species.

'Man is born free and is immediately reduced to slavery by society', said Jean Jacques Rousseau in *Du Contrat Social.* In human infants, the feeding pattern, amounts and nature of foods eaten are mainly imposed by the mother, who is guided by sociocultural uses or medical prescriptions. However, human babies attest by their violent cries to the presumed intensity of their hunger and perhaps to the inadequacy of this adult regulation regarding their craving for calories and material for growth. Some time ago, observations were made on babies whose mothers were asked to give the breast on demand. Despite some discrepancies between various reports it appears that babies asked 10 to 11 times per day from around 2–7 days after birth. After three months, babies stabilized to 5–6 meals distributed in a diurnal pattern (Symparian & MacLendon, 1945). In some reports these meals would be irregularly separated by 60 to 90 min of no-feeding. In other reports a 4 h cycle would be exhibited, with no relation between the size of each intake and previous or subsequent intervals of no-feeding (Morath, 1974). As does the rat, babies at an early stage revealed preferences for sweet and salty solutions, and rejection of bitter-tasting compounds (Desor *et al.*, 1973; Nowlis, 1973; Steiner, 1973; Lipsitt, 1977). Many studies (which cannot be analysed here) emphasize the prominent role, after weaning, of the sociocultural environment in teaching eating and this is concurrent with the physiological learning, until adulthood, of all types of behaviour associated with eating.

8 Brain mechanisms of hunger arousal and meal initiation

Data examined in Chapter 3 led to the conclusion that an imbalance between the tissue glucose uptake and the input of glucose into the blood brings about a systemic stimulus of hunger arousal. What is the target of this stimulus? In addition, it was suggested that this hunger signal exerts a permissive action on the sensory mediation of eating. How and where does brain processing of oro-sensory activities of foods occur and interact with the systemic humoral signal?

The lateral hypothalamic feeding system

A large number of experimental studies have identified the lateral hypothalamic (LH) area and its neural connections with other structures, particularly with the amygdala and the frontal cortex, as critical loci of interaction between the blood-borne signals and sensory input which govern feeding. Evidence for an LH feeding system is provided by neuroanatomical studies, by the effects of electrical stimulation and self-stimulation and of electrolytic and chemical lesions and, conclusively, by electrophysiological recordings.

Neuroanatomical evidence

Both the LH and amygdala are sites of convergence of external and internal ascending neural pathways. Gustatory afferent pathways terminate in the LH, in the basolateral nuclei of the amygdala and in the substancia innominata through the nucleus of the tractus solitarius (NTS) and the pontine relay (Norgren, 1976). Direct monosynaptic projections from the olfactory bulb to the anterior amygdala and from the piriform cortex of the LH are well documented (MacLeod, 1971; Kogure *et al.*, 1980). Responses to visual stimuli can also be recorded in the LH and amygdala (Riley *et al.*, 1981). Vagal and parasympathetic afferent pathways from the gastrointestinal tract and from the liver overlap the gustatory pathways and, via the dorsomotor nucleus of the vagus, the NTS and reach, on the one hand, via the parabrachial nuclei, the periventricular nuclei of the LH (Novin *et al.*, 1981) and, on the other hand, the basolateral nuclei of the amygdala (Ricardo & Koh, 1978). Sympathetic gastrointestinal and splanchnic afferent pathways, as yet poorly recognized, seem to terminate in the medial hypothalamus.

So the lateral and medial hypothalamus might be parasympathetic and sympathetic projection sites, respectively. These visceral limbic and hypothalamic connections are bi-directional. Descending parasympathetic pathways from the LH and basolateral amygdala (Saper *et al.*, 1976) involve the gastrointestinal tract and the pancreas and are in action in the neural control of gastric secretion and of the endocrine pancreas (Rogers *et al.*, 1980; Powley & Laughton, 1981). As suggested by electrical stimulations and lesions, sympathetic descending pathways from the medial hypothalamus are involved in the adrenergic control of the adrenal medulla and in the adrenergic innervation of white and brown adipose tissues, of alpha and beta cells of the pancreas, and of the liver. Connections have been recognized between the LH and the frontal cortex (Yamamoto & Shibata, 1979). Olfactory impulses via amygdaloid nuclei have also been recorded in the orbitofrontal cortex (Tanabe *et al.*, 1975). Finally, the LH is closely linked with the limbic structure by the basolateral amygdalo-fugal pathway and the stria terminalis. Fibres passing the LH may be neurochemically identified as components of the dopaminergic nigrostriatal bundle, while the other fibre system of this area belongs to the medial forebrain bundle and has multiple connections with other hypothalamic nuclei.

Electrical and chemical stimulation of the LH

The stimulation of feeding by electrical stimulation through electrodes implanted in the LH was demonstrated for the first time by Delgado & Anand (1953). Since then, electrical stimulation of any other part of the brain has been shown to provide the same effect. During daily sessions of two hours' duration stimulus-bound feeding (SBF) can induce a 2.5-fold increase over the previous control intake in rats. As mentioned in Chapter 6, this stimulation-induced hyperphagia, repeated daily, leads to obesity (Steinbaum & Miller, 1965).

There is much evidence that the state induced by LH electrical stimulation does not differ from that which underlies the normal food intake (Coons & Cruce, 1968). It has already been noted (p. 41) that the concentration of quinine in the diet, necessary to block SBF, increases as a function of current intensity just as it does as a function of food deprivation (Tenen & Miller, 1964). However, electrical stimulation of the LH may elicit drinking, gnawing or even sexual activity instead of feeding. The response obtained initially can be modified. If feeding rather than drinking is the initial response, repeated stimulation in rats presented with water may induce stimulus-bound drinking (Valenstein *et al.*, 1968; Valenstein & Cox, 1970). It was demonstrated that the initial response is maintained as long as sensory stimuli relevant to a particular behaviour are present (Huang & Mogenson, 1972; Watson *et al.*, 1979). With a particular electrode placement feeding is elicited in the presence of foods and drinking in the presence of water. This seems to indicate that electrical stimulation of the LH activates overlapping neuronal systems involved in thirst, hunger and so on, and that a particular response prevails when the relevant sensory input is involved (Wise, 1968). Results of self-stimulation of the LH further support this conclusion.

LH electrical stimulation produces hyperglycaemia, even though stimulation with the electrode in the same place does not induce feeding (Booth *et al.*, 1969). When feeding is elicited and during the stimulated meal, there is hyperinsulinaemia associated with hyperglycaemia Steffens, 1975). LH stimulation at a low current intensity increases the metabolic rate (O_2 consumption) transiently (Atrens *et al.*, 1985). It does not induce lipolysis (Shimazu, 1981). Thus no indication exists that the LH is involved in a neuronal control of blood glucose level or of body fats.

Chemical stimulation in the same area supports the same conclusion. If we look at the efficiency of a local injection of the glucose analogue, 2-deoxy-glucose (2-DG), we find that the 2-DG is either without effect on feeding (Miselis & Epstein, 1975; Berthoud & Mogenson, 1977) or stimulates it in fed rats (Balagura & Kanner, 1971), and in rabbits (Gonzalez & Novin, 1974). A small local infusion of glucose blocks the feeding response to peripherally administered insulin (Booth, 1968). This is confirmation that the true stimulus of insulin-induced feeding is the resulting glucopenia and suggests in addition that this glucopenia has its effect at the level of the LH.

Rats press a lever to obtain an electrical stimulation from electrodes placed in the LH and other parts of the brain. This well-documented intracranial self-stimulation (ICSS), when elicited from the LH (LH self-stimulation), provides important information about the feeding system. The rate of LH self-stimulation in rats increases with food deprivation. This increase is related more to weight loss induced by food deprivation than to the duration of the food withdrawal (Margules & Olds, 1962; Blundell & Herberg, 1968). This effect of food deprivation is observed only when stimulation by an electrode in the LH also induces SBF (Goldstein *et al.*, 1970). LH self-stimulation is also induced by peripherally injected insulin (Balagura & Hoebel, 1967) and in rats with ventromedial hypothalamic lesion, which (as will be seen later) is strongly hyperinsulinic (Hoebel & Teitelbaum, 1962). Except in extreme cases reported by Spies (1965), rats continue to eat normally when self-stimulation is available. Clearly the presence of food as a sensory stimulus and the degree of 'palatability' of the food act as co-reinforcers of self-stimulation (Coons & Cruce, 1968; Poschel, 1968; Mendelson, (1969). Only LH self-stimulation can be modulated by the hunger state and by food-related stimuli. This argues in favour of the additivity of the two sources of the stimulation to eat in promoting the self-stimulation and in turn further supports the notion that electrical stimulation of the LH reproduces a state identical with that initiating and rewarding oral intake.

In the absence of a concurrent food intake, a high rate of ICSS persists for a long time and as such resembles sham-feeding; for, as the latter ceases on manipulation of the gastrointestinal and systemic satiation mechanism, LH self-stimulation is suppressed or stopped by the various conditions which normally induce satiation or are conducive to persistent satiety. Gastric distension, gastric loads, glucagon administration and an experimentally induced obesity suppress LH self-stimulation (Wilkinson & Peele, 1962; Balagura & Hoebel, 1967; Balagura, 1968; Hoebel, 1968; Hoebel & Thompson, 1969; MacNeil, 1974).

LH lesion

Anand & Brobeck (1951) found that a bilateral electrolytic lesion of the area of the brain which could elicit feeding when electrically stimulated produced an aphagia–adipsia syndrome. Administration of 6-hydroxydopamine in the same region or in the striatum (Ungerstedt, 1971) produces a syndrome similar but not identical with that produced by electrolytic destruction. The syndrome can be reproduced after a local injection of kainic acid which induces a lesion limited to the cell bodies (Stricker *et al.*, 1978; Grossman & Grossman, 1982). Lesions of other structures connected to the LH (medial forebrain bundle, zona incerta, amygdala, frontal cortex and midbrain tegmentum) also induce aphagia or hypophagia (Le Magnen, 1983). Some of the effects of these lesions, e.g. of the orbito-frontal cortex, are in addition to those of the LH lesion.

The lesion-induced aphagia is reversible. After total aphagia or adipsia lasting 8–20 days or more, according to the placement and the extension of the lesion, rats recovered progressively their pre-operative intakes of food and water (Teitelbaum & Epstein, 1962; Teitelbaum & Stellar, 1954). After recovery from aphagia, rats show almost normal responses to homeostatic challenges, but responses to physiological stimuli of thirst and water intake are impaired. Recovered rats respond normally to food deprivation. However, the intake-promoting effects of insulin and glucose analogues are not restored (Epstein & Teitelbaum,

1967; Wayner *et al.*, 1971). This provides evidence that the insulin and 2-DG-elicited feeding responses are closely dependent on the destroyed LH sites and are not taken over by the system responsible for recovery of normal feeding and of responses to food deprivation.

After forced weight gain or forced weight loss, LH-lesioned rats which have recovered return to a stable (defended) body weight after a transient hypophagia or hyperphagia, as do the control rats. However, this defended body weight is 10 to 15% lower than that in unoperated controls. This fact has been presented as evidence for a role of the LH in the regulation of body fats (Keesey *et al.*, 1976). However, there is nothing to indicate that the LH is involved in this regulation, that is to say in an active correction of weight gain by a subsequent weight loss and vice versa. The lower body weight chronically maintained as thus defended in this experiment is presumably due to a new equilibrium between hormonal and neuronal factors controlling, respectively, lipogenesis and lipolysis. Other data suggest that a chronic hyperactivity of adrenal medulla (Opsahl, 1977), in addition to a chronic hyperinsulinism (Steffens & Lotter, 1979) could be responsible for this new equilibrium and therefore for the new level of defended body weight.

Effects induced by lesion, as those of electrical stimulation, suggest that the LH is not involved directly in the regulation of blood glucose level. After LH lesion, insulin and 2-DG, as mentioned above, no longer activate feeding; but their respective hypoglycaemic and hyperglycaemic effects are maintained (Panerai *et al.*, 1975).

The respective contributions of sensorimotor disturbances and of a loss of responsiveness to the systemic stimulus of hunger as causes of aphagia have been questioned. Sensorimotor disturbances are obvious consequences of the lesion (Marshall *et al.*, 1971). After a unilateral lesion, rats exhibit impaired responsiveness to olfactory, tactile and painful stimulations in the contralateral field. A role of 'sensory neglect' in aphagia is strikingly illustrated when two food-cups are placed in front of a lesioned rat: it is aphagic only with regard to the food cup placed contralaterally to the unilateral lesion (Marshall & Teitelbaum, 1974). As in aphagia, the rat can recover from the sensorimotor impairment, but without showing a clear temporal correlation between the regression of the two symptoms.

Attempts to elucidate the role of this sensorimotor impairment have brought conflicting results. Rats trained to press a lever for food continued to respond after the lesion but did not eat the food they earned. When trained to press for a gastric delivery of a liquid food, rats also continue to press after the lesion. In so doing they proved that they are still able to eat but eating by oral intake was impaired as a result of sensorimotor disturbance (Baillie & Morrison, 1963). In another study the lever pressing of LH-lesioned rats for a gastric delivery of foods was not observed and the conclusion was drawn that the lesion abolishes the systemic stimulation to eat (Rodgers *et al.*, 1965). This suggests that the LH lesion may impair both the sensory and the systemic components because they converge and interact in the same region. However, recent experiments indicate a topographical dissociation between lesion sites which leads to aphagia predominantly determined by either the sensorimotor or the metabolic component, or by both, according to the extent of the lesion. Sensorimotor disturbance would prevail after lateral and postero-dorsal lesions. Unresponsiveness to systemic stimuli would be predominant after lesion of antero-ventral sites in which (as will be seen below) glucose-sensitive sites have previously been located (Schallert & Whishaw, 1978).

The brain target of the systemic and blood-borne hunger signal

In 1971, Russek put forward a theory according to which hepatic gluco-receptors would be the detectors of the nutritional imbalance which stimulates feeding and causes the adjustment to this imbalance. The recording of electrical discharges in the hepatic branch of the vagus which vary with the concentration of various sugars injected into the portal vein seemed to support this (Niijima, 1975). However, complete denervation of the liver including the section of the hepatic branch of the vagus nerve did not alter the normal food intake and its meal patterning (Louis-Sylvestre *et al.*, 1980; Bellinger & Williams, 1981), thus disproving the theory. A porto-caval anastomosis also did not significantly impair the normal feeding pattern in rats (Louis-Sylvestre *et al.*, 1979).

Some time ago it was suggested that the LH was the target of a blood-borne signal giving rise to hunger arousal and feeding. Initially, it was found that multi-unit neural discharges in the LH were elevated in food-deprived animals (Anand *et al.*, 1962). More recently it has been shown that activation of single units in the LH by food-related stimuli occurs in hungry animals only (Rolls *et al.*, 1976). Single-unit discharges acutely or chronically recorded in the rat LH are inhibited while the rat is eating the food (Hamburg, 1971; Ono *et al.*, 1981).

Electrophysiological investigations have also revealed that some of the LH neurones behave in a glucose-sensitive manner, i.e. their firing rate may be modified by minimal variations in the glucose concentration in their environment (Oomura, 1976). Local iontophoretic application of glucose either activates or inhibits the firing of a majority of the neurones in the ventral portion of the LH. Adding insulin to the local infusion of glucose did not change the response to glucose, but insulin alone activates LH neurones in a dose-dependent manner. Free fatty acids (FFA), added to the glucose do not reverse the glucose-induced neuronal inhibition. Strong support for the existence of such glucose-sensitive neurones in the LH was provided by the finding that a gold thioglucose (GTG) implant in the LH produces aphagia (Smith, 1972). GTG induces a necrosis in the specific affinity of cells for glucose. This GTG-induced necrosis in the LH is identical in diabetic and in normal rats. The uptake or binding of the neurotoxin by LH cells is not therefore insulin dependent.

The recovery of feeding and of homeostatic challenges in LH-lesioned rats may cast doubt on the role of such glucose-sensitive neurones in feeding. Possibly the recovery is due to the glucose-sensitive specific chemosensors located in the LH being replaced by a more diffuse system which is also dependent on the glucose supply. The fact that cortical spreading depression[1] reinstates aphagia in LH-recovered rats suggests that the neocortex could be the diffuse system (Teitelbaum & Cytawa, 1965).

Whether these brain chemosensors governing feeding have a specific response to glucopenia has been also questioned. Theories have been advanced which suggest that the availability not only of glucose but also of other mutually interchangeable energy metabolites (FFA, proteins, ketone bodies) may be involved (Booth, 1972*b*). This seems unlikely at first because peripheral glucopenia has only been correlated with feeding events and therefore can be only a candidate for the systemic stimulus to the brain. On the other hand, energy metabolites are not interchangeable as substrates of energy metabolism in brain cells. Ketone bodies, FFA and local break-down products of amino-acids can only substitute for glucose in 40 to 50% of the energy metabolism and even this occurs only after a delay of the order of 24 h under fasting (Smith *et al.*, 1969). Chronic brain feeding provides evidence for this lack of equivalence of metabolites as reflected in changes in food intake. An intraventricular infusion of glucose reduces food intake (Herberg, 1960); glycerol reduces it more persistently; and ketone bodies are without effect. The relative intake-reducing effects of glucose and glycerol are different by night and by day (Davis *et al.*, 1981). The correlation between lipolysis and hypophagia provides strong evidence that peripheral oxidation of FFA, reducing glucose utilization, retards

[1] Functional dysconnection of the cortex due to KCl topical administration.

the occurrence of glucose shortage and the resultant brain activity which stimulates feeding (Balasse & Neef, 1974).

The exact nature of the glucose chemoreception of responsive neurones in the LH is absolutely unknown. We do not know whether their stimulation is due to a suppression of glycolysis and hence of the production of high energy substrates, or to a deficit of the glucose transport across the cell membrane, or to a reduction in membrane glucose-binding sites as has been suggested for gustatory cells and the beta cells of the pancreas (Woods & McKay, 1978). Alloxan intraventricularly administered abolishes 2-DG-stimulated but not normal feeding (Woods & McKay, 1978). This is again contrary to the notion that feeding stimulation by 2-DG reproduces the normal condition of the metabolic stimulation to eat.

Brain mechanisms of palatability of foods

As previously stated, meal initiation and the initial rate of eating an accepted food result from the additivity of the respective strengths of systemic stimulation and of sensory stimulation (palatability of food). The brain mechanisms involved in the manifestation and acquisition of the differential palatability of foods and of the adjustment of these mechanisms to different nutritive properties have been investigated.

Manifestations of palatability

The amygdala and its connections with the LH seem to be the main areas of the brain involved in the manifestation of genetically determined and acquired food palatability. It has been demonstrated that lesions of various nuclei of the amygdala, alter the selective response to foods and fluids which is based upon their different sensory properties (Fonberg & Sychowa, 1968; Rolls & Rolls, 1973). Lesion of the basolateral nuclei eliminates the neophobic response to a novel food item. Lesion of the adjacent lateral nuclei enhances the manifestation of a high palatability of a sweet-tasting solution as soon as it is first presented (Kolakowska *et al.*, 1984). However, active rejection of food in spontaneous or acquired true aversion is dependent on other structures. Lesions in the septum and in area postrema (involved in nausea and vomiting) alter these defensive reactions to food items (Ashe & Nachman, 1980). Integrity of the gustatory neocortex is not necessary for the manifestation of these rejections as indicated by the facial expressions of rats after infusion of a quinine solution in the mouth (Grill & Norgren, 1978*a*).

The sensory food reward

This sensory support of eating or palatability may be called 'the sensory reward of the eating response'. Might 'brain rewarding systems', identified by the electrical self-stimulation, be involved in food reward and therefore in palatability? This idea is strongly supported by the data on the electrical self-stimulation of the LH described above. Also, two strains of rat have been selected for high and low LH self-stimulation. Saccharin intake was shown to be higher in rats exhibiting the high rate of self-stimulation than in low self-stimulators (Ganchrow *et al.*, 1981).

There is much controversy surrounding the neurochemical substrates of the so-called 'brain rewarding systems' considered to be sites of reinforcement of ICSS. The prevailing opinion is that they are dopaminergic. Pimozide, which is a dopaminergic blocking agent, not only suppresses the ICSS but also the food reward of an instrumental response. This supports the idea that dopaminergic brain rewarding systems are involved in both electrical stimulation and food rewards (Wise *et al.*, 1978). However, Esposito & Kornetsky (1978) found evidence for an additional involvement of brain opiates in mechanisms of reinforcement of ICSS and natural behaviour. Thus, the new and important possibility of a role of brain opioids and opioreceptors in feeding and particularly in support of food palatability was raised.

Brain opiates and palatability of foods

In 1980, it was shown for the first time that injection of naloxone, an opiate antagonist, abolished the high intake of a saccharin or glucose solution, compared to water, and also exaggerated the low palatability of a quinine or an ethanol solution (Le Magnen *et al.*, 1980*b*). Independently, several other investigators showed that the same opiate antagonist reduced food intake in food-deprived and undeprived rats (Cooper, 1980; Morley *et al.*, 1983). This reducing effect was interpreted by Morley *et al.* to be due to a possible involvement of brain opioids in meal initiation. Specifically, kappa opioreceptors were suggested to be involved on the basis of pharmacological testing by agonists and antagonists of these opioreceptors (Morley *et al.*, 1983). Indeed food deprivation, and stress induced by 2-DG administration, have been shown to cause beta-endorphin release from the hypothalamus and a subsequent depletion of this neuropeptide in the area (Bodnar *et al.*, 1981). But a role of this opioid release in the initiation of feeding, i.e. in hunger arousal, is doubtful because release of encephalin, dynorphin or beta-endorphin bears no apparent relationship to the metabolic events which determine meal initiation.

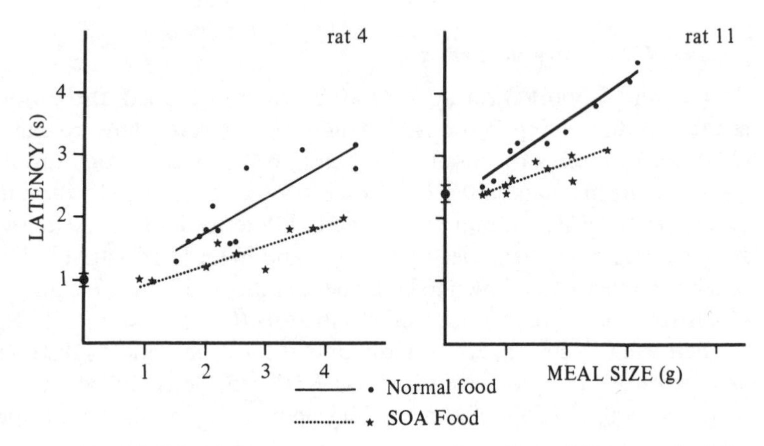

Fig. 8.1. Relation between amounts eaten of two foods of different palatability and the post-prandial response to pain in individual rats. Analgesia indicating brain opiate release is proportional to amounts eaten and higher with the high- than the low-palatability food. SOA = sucrose octo-acetate.

But there is support for a role of brain opioids in the sensory food reward. It has been demonstrated that the intake of a saccharin solution by rats induced a release of beta-endorphin and a consequent depletion in the hypothalamus (Dum *et al.*, 1983). The well-documented attenuation of nociception and pain by the induced release of brain opiates indicated that analgesia, as a measure of opioid release, could occur after a meal and could be related to the palatability of the food eaten. This hypothesis was confirmed by the finding of a 'post-prandial analgesia' (J. Louis-Sylvestre, M. Lagaillarde and J. Le Magnen, unpublished) (Fig. 8.1). Analgesia was tested by a modified tail-flick test in rats 30 min after the end of a spontaneous meal. A highly significant positive correlation was found between the amount eaten in successive test-meals and the level of analgesia induced by these meals. Furthermore, in the same rats and for the same amount of food eaten, the level of the induced analgesia was lower when a low palatability food was substituted for the high palatability familiar diet. Naloxone injected 15 min before the tail-flick test abolished the post-prandial analgesia and induced hyperalgesia.

Subsequent experiments have opened the way to a series of fascinating investigations which really lie outside the scope of feeding. It was reported that chronic stress in the mouse led to chronic release of brain opioids and could establish a state of morphine-like dependence suggested to be a dependence of the animal on its endogenous opiates (Christie & Chesher, 1982). It was hypothesized that a chronic intake of high palatability foods and the consequent hyperphagia could also induce a state of food addiction caused by a morphine-like dependence. This hypothesis has been confirmed (J. Louis-Sylvestre, M. Lagaillarde and J. Le Magnen, unpublished). Rats fed over a long period on a cafeteria regimen were challenged by a naloxone injection at the discontinuation of this cafeteria feeding. The animals exhibited various neurological signs of a precipitated morphine withdrawal syndrome similar to that displayed in a state of morphine dependence. Independently other investigators have shown that a chronic intake of sweet saccharin solutions in rats induced a tolerance to the analgesic effect of exogenous morphine, which was suggested to be a cross-tolerance to endogenous opiates repeatedly released by the sensory food reward (Lieblich *et al.*, 1983).

The brain mechanism of palatability learning

What is the brain mechanism which causes modulation of food palatability as a result of learning?

Here again brain systems involved in the reinforcement of electrical self-stimulation could be involved. This is supported by experiments which show that the pairing of a saccharin intake with electrical stimulation of a site in the brain which causes rats to self-stimulate enhances the response to saccharin, in other words induces a 'conditioned taste preference'. The same electrical stimulation of brain rewarding structures antagonizes the conditioning of an aversion when the stimulation and LiCl administrations are paired simultaneously with the intake of a saccharin solution (Ettenberg & White, 1978; Ettenberg *et al.*, 1982). However, as in other types of learning, it is difficult to localize the learning of palatability to a particular brain structure. It is likely that a complex neural circuitry between sensorimotor pathways is involved.

There have been a considerable number of investigations into the effect of various brain lesions on the conditioning of taste aversion. Lesions in the LH, area postrema and gustatory neocortex have been shown to alter or to abolish the capacity of rats to associate the response of a taste stimulus with a post-ingestive toxicosis (Ashe & Nachman, 1980). However, attention has been focussed on a particular role, at least as an important relay, of the basolateral and lateral nuclei of the amygdala. After a lesion of the basolateral nuclei, the conditioning of a lowered palatability of a saccharin solution by association with LiCl, is abolished. Furthermore, the same lesion prevents the progressive augmentation of the intake of a briefly presented sucrose solution, which indicates a conditioning of taste preference (Kolakowska, 1984). Thus,

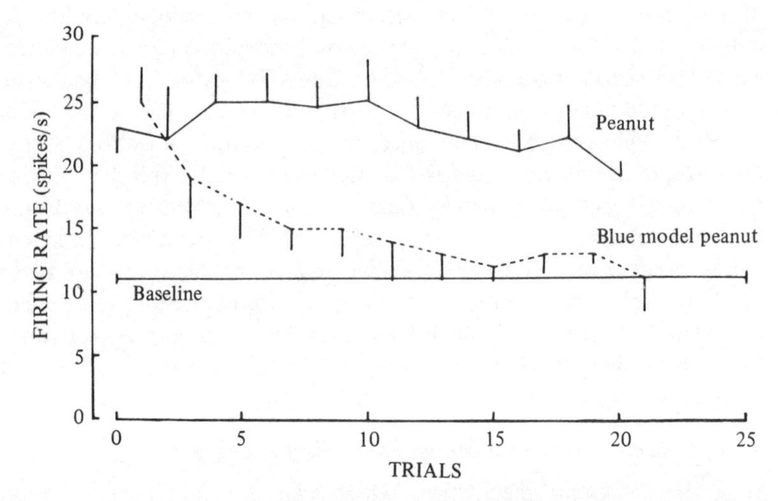

Fig. 8.2. Firing rate of a lateral hypothalamic unit during the learning of visual discrimination. On alternate trials the monkey saw for 10 s, and was then fed, a real peanut, or he saw for 10 s an artificial peanut made of blue modelling clay (blue model peanut). Initially the response to both stimuli was an increase of the firing rate from the baseline spontaneous rate of 11 spikes. During the next few trials the response to the blue model peanut became smaller and the response to the real peanut remained approximately constant.

lesion of the same structure impairs the conditioning of both an enhancement and a lowering of food palatability. This again suggests that palatability is a continuum.

Electrophysiological recordings identify palatability learning at the neuronal level and the role of neurones in a central processing of sensory afferent information. In the monkey, the firing rate of a single unit in the LH is altered by the presentation of a food object. The same unit, initially identically responsive to a real peanut and to an artificial peanut, responds only to the former when both have been previously experienced by the monkey (Fig. 8.2) (Mora *et al.*, 1976). Similarly, the multi-unit discharges of the mitral cell layer of the olfactory bulb are enhanced in hungry rats, only when the animals are stimulated by a food-related odour. Cells not responsive to a food-unrelated odour become responsive to this odour when it is previously associated to the familiar food (Pager *et al.*, 1972). The response recorded in hungry rats to the food-related odour disappears when the odour is associated with a LiCl injection and hence has become relatively aversive (Pager & Royet, 1976). The same suppression of or alteration in the pattern of single-unit discharges in the nucleus of the tractus solitarius evoked by a sweet solution has been found after the sweet solution has been made aversive by toxicosis (E. M. Scott and F. C. T. Chang, unpublished). Thus palatability learning seems to modify, presumably through the centripedal pathways, the sensory input to critical sites of sensorimotor connections.

9 Brain mechanisms of satiation and meal size

What is the neuronal mechanism by which food passing the mouth and entering the stomach and/or the intestine acts as a negative feedback to counteract the initial stimulation to eat until the onset of satiety is reached? In other words, what is the brain mechanism which determines the size of meals?

Satiety is established at the end of a meal by the satiation process during the meal. The persistence of this state of satiety involves no other mechanism than that involved in its ending by the initiation of a new meal. Thus it is meaningless to suggest a distinction between a hypothetical systemic mechanism of satiety and another systemic mechanism of meal initiation. In fact, there is no evidence for a specific systemic stimulus, including hyperglycaemia or other blood-borne signal, as a satiety signal. Therefore it was not reasonable to look for and to assume the existence of a brain target of this putative signal and to support the notion of a 'satiety centre' distinct from a 'feeding centre'.

Non-involvement of the medial hypothalamus

For a long time (as a result of historical error) the ventromedial hypothalamic nuclei (VMN) have been considered to be this nebulous centre of satiety. The concept was based upon the induction of hyperphagia and obesity by lesion of these hypothalamic nuclei. As it will be detailed in the next chapter, it has now been fully demonstrated that this hyperphagia is not due to a loss of the capacity of lesioned animals to be satiated by eating the food. On the contrary, after the VMN lesion, both the mechanism of meal initiation, which limits the duration of satiety, and that of satiation, which limits meal sizes, are unaltered (Panksepp, 1971; Smutz *et al.*, 1975).

It has been classically demonstrated that electrical stimulation of the VMN blocks the feeding response in hungry animals. However, inhibition is observed during stimulation and persists for only 10 min after its termination. The reduced intake is compensated for by a subsequent increase in intake. In rats habitually fed for 6 h daily, repeated stimulation does not lead to a reduction in intake over the 6 h period (Smith, 1958). Furthermore, it has recently been shown that inhibition is dependent on the placing of the electrode and on current intensity.

Low-intensity stimulation through an electrode placed in the laterodorsal part of the VMN increases rather than decreases the feeding of a food-deprived animal. With the same emplacement, inhibition is obtained with stimulation of a higher intensity (Davies *et al.*, 1974).

The postulate (based upon the inhibitory effect of electrical stimulation) that the ventromedial hypothalamus (VMH) acts as satiety centre through neural inhibition of lateral hypothalamic (LH) areas has been ruled out. Rats may learn to press a lever to escape programmed stimulation of the VMH. If VMN were involved in satiety, food deprivation should decrease this escape response; this is not the case (Goldstein & Ripley, 1976; Horell & Redgrave, 1976). The inhibitory effect of VMN electrical stimulation persists after a parasagittal knife cut is made between the medial hypothalamus and the LH and is still observed in rats which have recovered from LH lesions (Sclafani & Berner, 1977).

Role of the LH and connected structures in the satiation process

All the data now available demonstrate that the pre-absorptive satiation process is due to the projection of both sensory and sympathetic/parasympathetic pathways coming from the alimentary canal to the system involved in the onset of feeding, namely the LH; in other words, the engagement, maintenance and termination of eating are dependent on the same brain feeding system involving the LH and connected structures. In rats with LH lesion, after the recovery of a normal daily food intake, the disappearance of the prandial periodicity (the capacity of animal to eat in intermittent feeding episodes) always follows. The post-operative rat becomes a nibbler, eating almost continuously at night in 2–3 min bouts separated by pauses of some minute duration which are occupied by drinking. The circadian periodicity with the higher intake at night is maintained (Kissileff, 1970; Rowland, 1977) (Fig. 9.1). This disappearance of meal eating is also observed after other lesions: bilateral ablation of olfactory bulb, and lesions of the medial forebrain bundle and of the stria terminalis (Larue & Le Magnen, 1972, 1973). So these various limbic and hypothalamic structures and their connections may be involved in the mechanism which determines the termination of a meal after a given amount of a particular food has been eaten.

The involvement of the LH and its upstream pathways in the satiating process has been confirmed by electrophysiological recordings in the central nervous system. In the monkey, the firing rate of a single unit in the LH, inhibited by hunger and the sight of a food object, is

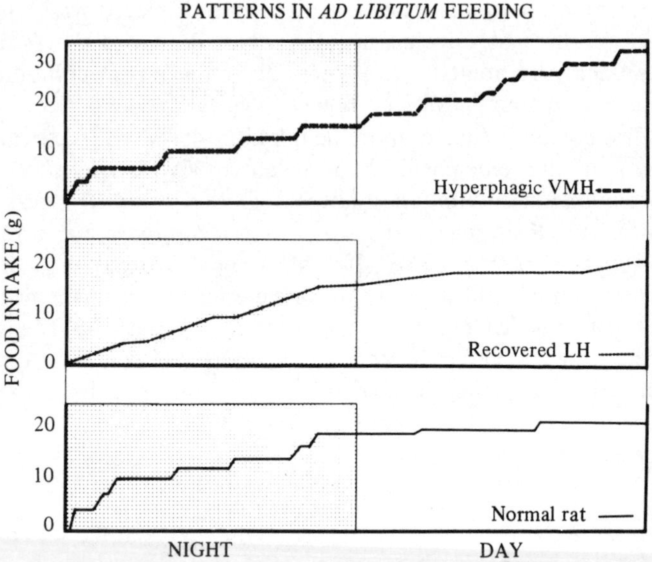

Fig. 9.1. Comparison of the daily recorded meal patterns of normal, ventromedial hypothalamus (VMH)-lesioned and lateral hypothalamus (LH)-lesioned rats. In VMH-lesioned rats, prandial periodicity is preserved and circadian periodicity abolished. In recovered LH-lesioned rats, prandial periodicity is abolished and the circadian one is maintained.

progressively resumed by eating the food (Rolls & Rolls, 1981) (Fig. 9.2). The sensory specificity of satiation is manifested by the resumed activity of the neurone when a new and different food is offered to a monkey satiated by the first one (Fig. 9.3). The satiation process is manifested in the first central relay of olfactory and gustatory afferent pathways. At the end of a meal or when a balloon is inflated in the stomach, electrical discharges in the mitral cell layer of the olfactory bulb are inhibited even in response to food-relevant stimuli (Pager *et al.*, 1972; Chaput & Holley, 1976) (Fig. 9.4). The same shift of responsiveness is observed, in respect to satiety, in the nucleus of the tractus solitarius (Glenn & Erickson, 1976). This unresponsiveness to the peripheral gustatory stimulation in satiation is observed when a sweet-tasting stimulus but not when a bitter-tasting stimulus is applied to the tongue. In the hypothalamic endings of gustatory pathways the same sensory-specific and food-related modulation of neuronal activity is recorded (Aleksanyan *et al.*, 1976).

What afferent pathways to the brain convey information from the alimentary canal to counteract the initial stimulation to eat which has been given by eating the food?

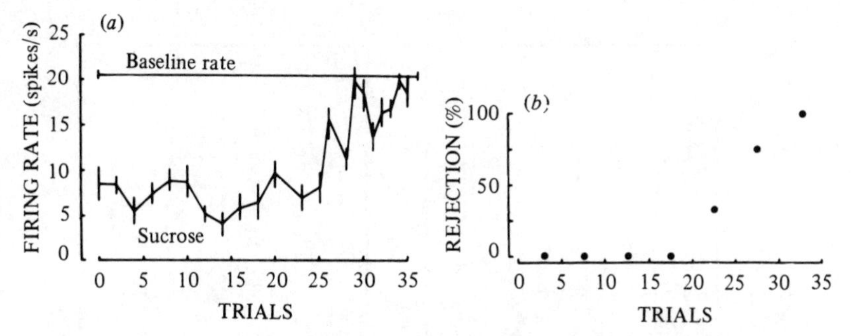

Fig. 9.2. (*a*) Effect of the sight of a syringe from which the squirrel monkey was fed 2 ml of 20% sucrose solution on the firing rate of a single hypothalamic unit. The rate decreased below the spontaneous baseline rate (the mean and SEM are shown) at the start of the experiment when the monkey was hungry but not at the end of the experiment when the monkey was satiated. The firing rate and its standard error were measured over a 5 s period during which the monkey was looking at the syringe. (*b*) Percentage of rejections, i.e. level of satiation as a function of successive intakes.

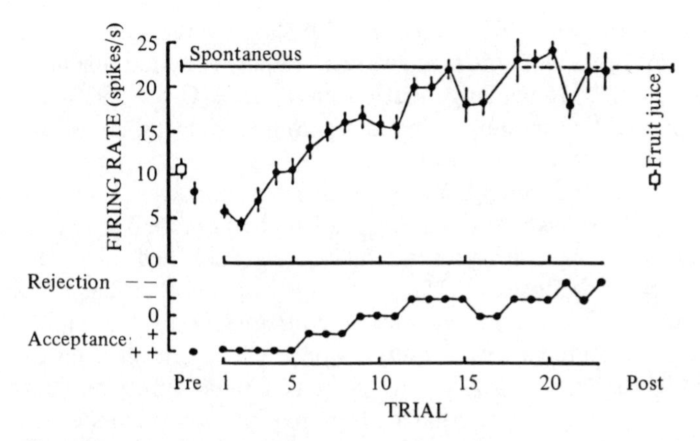

Fig. 9.3. Sensory-specific decrease in the responsiveness of a lateral hypothalamic neurone produced by feeding.

The participation of oro-sensory factors in the sensory-specific satiation process involves the same sensory afferent pathways as are involved in the initial facilitation. The transmission of information through the vagus from the chemosensors identified in the duodenum may be involved in the entero-insular axis. Through an afferent–efferent vagal loop these intestinal chemoreceptors may be responsible for the peak of insulin release at the transition between the pre- and post-absorptive phases which, in turn, may be effective in the final step of

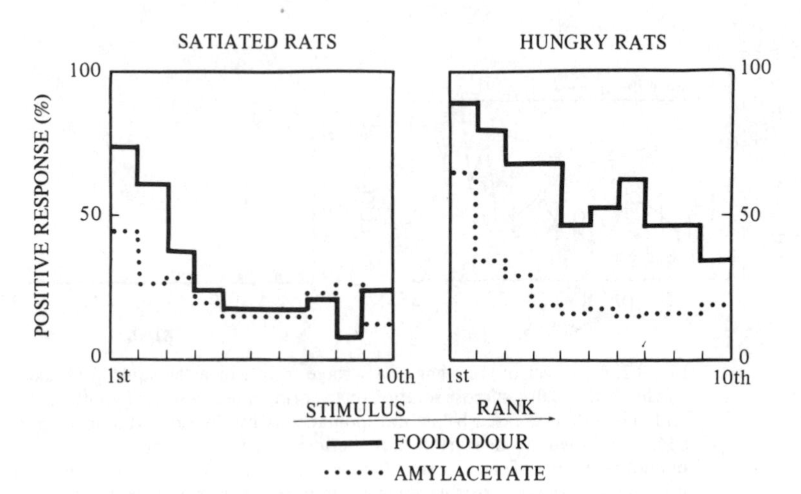

Fig. 9.4. Hunger elevates the percentage of positive electrical responses in the mitral cell layer of the olfactory bulb when rats are stimulated by the odour of their food and not when stimulated by a control odour (amylacetate).

satiation (Mei, 1978; Mei *et al.*, 1981). However, afferent fibres of the vagus nerve from the stomach seem to play a limited role in the normal satiation by food filling the stomach (Kraly & Gibbs, 1980). If they were dominant, a subdiaphragmatic vagotomy should induce meal overeating comparable to sham-feeding or to the fragmented pattern observed after LH and other brain lesions. But this is not the case. Vagotomized rats display a typical meal pattern of smaller and more frequent meals. The circadian periodicity is attenuated or abolished (Snowdon, 1969; Snowdon & Epstein, 1970; Louis-Sylvestre, 1978*a*). The reduced meal size has been interpreted as a consequence of the rapid stomach emptying induced by vagotomy and by a conditioned taste aversion, induced by the aversive effects of this rapid emptying (Bernstein & Goehler, 1983). Sympathetic fibres may be involved in the transmission of gastric cues (other than distension) which make the main contribution to the process. Their route to the brain and probably to the forebrain is unknown (Deutsch *et al.*, 1980).

The short-term determinants of meal size, i.e. of a filling of the gastrointestinal store by various amounts of foods, do not ultimately play a major role in feeding, i.e. the adjustment of the meal-to-meal intake to metabolic demands. Through the balance between the different initial palatabilities of the food and the post-oral satiating process, and through the effect of variety in mixed meals, these meal sizes vary as widely with rats as they do with humans. The bigger or smaller meals are corrected promptly by the adjustment of the post-prandial duration of satiety and therefore by the mechanism of initiation. They are more slowly adapted to body homeostasis by the conditioning or re-conditioning of the palatability of the food. In these two regulatory adjustments, the diencephalic limbic system is definitely involved. A third control system is the brain liporegulatory mechanism and its indirect impact on feeding.

10 Brain mechanisms of body energy balance and of maintenance of fat mass

The medium- or long-term balance between the mean rate of energy expenditure or retention and the mean rate of energy intake described in Chapter 5 does not involve centrally governed mechanisms other than those discussed in Chapters 8 and 9. Despite asynchronous fluctuations in obligatory energy losses and in energy gains by food intake, a regulatory adjustment of the gain to the loss is performed by the brain mechanisms involved in meal initiation (or hunger) and the palatability of foods (or sensory-specific appetites).

Short- or eventually long-term imbalances between energy outflow and inflow are buffered by mobilization or storage of body fats which thereby act as the body energy store and energy 'ballast'. But, as described in Chapter 6, an autonomous liporegulatory mechanism insures a limit of both the filling and the depletion of the fat reserves. Through this regulatory mechanism an over-repletion of the store can be corrected by subsequent lipolysis, while an excessive depletion can be corrected by subsequent lipogenesis. In addition, it was shown that this regulation of body fat mass appears to be a primary mechanism which indirectly affects the feeding mechanism proper by adding or subtracting metabolic fuel to the balance between lean tissue energy utilization and energy intake. This liporegulatory mechanism is governed by the central nervous system and is neuronally controlled.

The hypothalamic regulation of body fat mass

In 1940, Hetherington & Ranson discovered that the electrolytic lesion of the ventromedial nuclei (VMN) of the hypothalamus in rats induced a persistent hyperphagia leading to obesity. Since this finding, several thousand articles have been devoted to this phenomenon. The hypothalamic-lesion induced hyperphagia and obesity occur in various other species. Lesion by 6-hydroxydopamine of the ascending noradrenergic bundle also induces hyperphagia and obesity, but they differ in some aspects from the syndrome produced by the VMN electrolytic lesion (Ahlskog *et al.*, 1975). Parasagittal and coronal knife cuts reproduce the syndrome presumably by interrupting connections with the VMN. The paraventricular nucleus seems to be a critical focus of the involved neuronal pathway. Lesion of this area produces hyperphagia and obesity similar to those induced by the VMN damage (Gold *et al.*, 1977; Leibowitz *et al.*, 1981; Aravich & Sclafani, 1983), and because of the hyperphagia this longitudinal pathway was viewed as a component of the satiety mechanism.

The gross consequences of the bilateral electrolytic lesions of VMN are as follows: Hyperphagia and a rapid weight gain appear after surgery, immediately after recovery from anaesthesia, when foods are available in excess. The hyperphagia and rapid weight gain are maintained for 3–4 weeks, a period termed 'the dynamic phase of obesity' (Brobeck, 1946). The daily weight gain then decreases progressively until a plateau is reached, while the daily food intake returns to or near the pre-operative level. This period is termed 'the static phase of obesity' (Brobeck, 1946). Under fasting, VMN-lesioned rats return to their pre-operative body weight or less. When re-fed, they display a new dynamic phase of hyperphagia and weight gain until they have recovered the original static level of obesity. When an overweight condition beyond this plateau of obesity is induced by protamine-zinc insulin treatment or by gavage, rats become hypophagic at the discontinuation of the treatment, losing the extra weight gained until they reach the plateau of obesity again (Hoebel & Teitelbaum, 1966). Thus, the new level of elevated body weight is defended.

The maximal level of obesity reached and maintained during the static phase represents a four- to five-fold increase or more of the body fat mass and a doubling or more of body weight. During the dynamic phase, the positive energy balance is represented by an increase in the daily food intake, reaching twice the pre-operative level. As mentioned earlier, the prandial periodicity is maintained, but the dark–light periodicity has disappeared. The normal nocturnal hyperphagia is exaggerated, with larger and more frequent meals and a 20–30% increase in the 12 h intake. The daytime intake increases two- or three-fold with a meal pattern and a 12 h cumulative intake which becomes identical with the nocturnal one (Fig. 9.1). The normal daytime hypophagia is therefore abolished (Teitelbaum & Campbell, 1956; Balagura & Devenport, 1970). Concurrently, the normal nocturnal weight gain and daytime proportional weight loss are replaced by a sustained 24 h weight gain reaching or exceeding 10 g per day. Likewise indices of the diurnal lipogenesis–lipolysis pattern and of its neuroendocrine control are suppressed or abolished. Although the O_2 consumption (metabolic rate) is not changed, the respiratory quotient indicates a permanent and active lipogenesis, substituted for lipolysis during the daytime (Le Magnen *et al.*, 1973) (Fig. 10.1). Basal and stimulated insulin release, glucose disappearance rate or glucose tolerance are augmented at night.

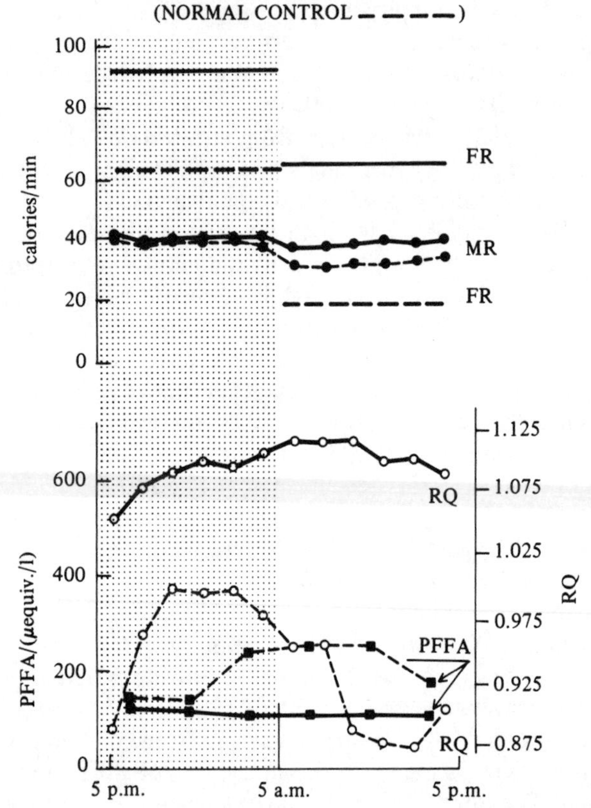

Fig. 10.1. Following VMH lesion, feeding rates (FR) become higher than metabolic rates (MR) throughout the 24 h. A permanently elevated respiratory quotient (RQ) indicates a resulting sustained fat synthesis which leads to obesity. PFFA = plasma free fatty acids.

This hyperinsulinism and high glucose tolerance are substituted for the normal opposite trend during daytime (Penicaud *et al.*, 1983). A continuous infusion of insulin reproduces exactly the hyperphagic and weight gain pattern in unlesioned rats. But contrary to the situation with intact rats, which correct the insulin-induced weight gain at the cessation of the treatment by a subsequent lipolysis and hypophagia, hypothalamic-lesioned rats maintain their hyperphagic-lipogenesis pattern until the new upper level of the static phase is achieved.

The main characteristics of the syndrome induced by the VMN lesion led to the following preliminary conclusions:

1. The liporegulatory mechanism by which an over-repletion of body fats is limited and corrected by a subsequent lipolysis and vice versa, is eliminated by the VMN lesion.
2. This mechanism is therefore dependent on an intact VMN.
3. The impairment of the induction of lipolysis is the main effect of the lesion and therefore this lipolysis is governed from the VMN by efferent neural pathways.
4. This impaired lipolysis is associated with an enhanced hyperinsulinic–lipogenesis pattern.
5. An induced imbalance between the sympathetic and parasympathetic innervation may be a main effect of the lesion.
6. Hyperphagia is a secondary effect of the diversion of metabolizable fuel to sustain fat deposition.

Further studies, as outlined below, have confirmed and extended these preliminary conclusions.

Effect of electrical stimulation of the VMN

Low intensity electrical stimulation of the VMN as well as of the dorsomedial nuclei and the arcuate nuclei produces a transient rise in blood glucose that seems to be due more to a splanchnic increase in hepatic glucose production than to adrenal catecholamine release. Stimulation of higher intensity leads to hyperglucagonaemia, which is dependent on neural afferents to the pancreas (Frohman & Bernardis, 1971). Finally, hypoinsulinaemia contributes to a sustained hyperglycaemia. This hypoinsulinaemia is abolished after adrenalectomy. The neuroendocrine pattern is similar to that induced by systemic and intraventricular administration of 2-DG. The hyperglycaemic–hypoinsulinic pattern during VMN stimulation is associated with decreased activity in the pancreatic branch of the vagus nerve (Oomura & Kita, 1981). In perfused rat pancreas, the VMN electrical stimulation modifies glucose-induced insulin release (Curry & Joy, 1974; Girardier *et al.*, 1976).

The most important point is that electrical stimulation of the VMN elicits lipolysis. In rats, stimulation is followed immediately by a rise of plasma glycerol and, in anaesthetized animals, by a rise in plasma free fatty acids. This lipolysis is not affected by removal of the adrenal medulla. Therefore it is not dependent on sympathetic efferent pathways to the adrenal medulla and hence on catecholamine release. Shimazu (1981) and Kumon *et al.* (1976) have suggested that lipolysis is stimulated by the VMN through sympathetic fibres innervating the adipose tissue. The above results also suggest that a sympathetic inhibition of insulin release and stimulation of glucagon release may also be involved. Electrical stimulation of the VMN reduces food intake, probably not because these nuclei are involved in satiety, but more likely because of the induction of lipolysis which in turn induces hypophagia, as it does during the daytime in normal rats.

VMN and LH relationships

A knife-cut between the VMN and the LH does not prevent the induction of hyperphagia by a second knife-cut medial to the first one or by electrolytic damage to the VMN (Sclafani *et al.*, 1974). This put paid to the idea that hyperphagia was due to the loss of an inhibitory effect of the VMN on the LH. Indeed a lesion in the LH prevents hyperphagia when a VMN lesion is performed subsequently, or abolishes hyperphagia when the VMN lesion is performed first (Anand *et al.*, 1955). But an intact LH is not necessary to achieve the effect of VMN destruction; all that is required is that the rat be able to eat. In rats recovered from LH aphagia, hyperphagia is obtained after a ventromedial lesion; it also reappears if the VMN lesion has been performed before the lateral hypothalamic lesion (Sclafani & Maul, 1974). Furthermore metabolic symptoms of ventromedial damage persist when aphagia is induced by a subsequent LH lesion. In rats lesioned at weaning, these metabolic symptoms without hyperphagia are maintained after a subsequent LH operation (Schnatz *et al.*, 1973).

Sustained lipogenesis and impaired lipolysis after the VMH lesion

The increase in lipogenesis in lesioned rats, substantiated by weight gain, can be assessed by the elevation of incorporation of injected [^{14}C]glucose or [^{14}C]acetate into body fats (Goldman *et al.*, 1972*b*; Løvø & Hustvedt, 1973). After some delay, the lipoprotein lipase activity (LPL) of adipocytes is augmented four-fold. This increase of LPL activity is identical with that developed in unoperated controls under exogenous insulin (Lowell *et al.*, 1980), and thus must be entirely dependent on hyperinsulinaemia.

In addition to hyperinsulinism as the major cause of the increased lipogenesis, evidence exists for an enhanced adipocyte responsiveness to insulin. Rats lesioned in the VMN at weaning and placed on a 3 h daily feeding schedule exhibit elevated adipocyte responsiveness to insulin *in vivo* and *in vitro*, although they are neither hyperphagic nor hyperinsulinic (Goldman *et al.*, 1972*a*; Goldman & Bernardis, 1974). The opposite, insulin resistance, is not a primary effect of the lesion. The resistance develops progressively in parallel with, and probably as an effect of, developing obesity (Han *et al.*, 1972; Le Marchand *et al.*, 1978). The static phase of obesity seems to result from a new balance between hyperinsulinaemia and the established insulin insensitivity and unresponsiveness of the hypertrophied adipocytes.

The impaired lipolysis has been confirmed *in vivo* and *in vitro* (Haessler & Crawford, 1967; May & Beaton, 1968). *In vitro*, the catecholamine-induced lipolysis of adipocytes from VMN-lesioned rats is suppressed (Kasemsri *et al.*, 1972). In the cold, during exercise and under fasting, lipolysis is lower and/or increases more slowly in lesioned rats than in unoperated controls (Nishizawa & Bray, 1978).

Role of hyperinsulinism and of sympathetic–parasympathetic imbalance

Rats with VMN lesion are strongly hyperinsulinic. The basal and food-stimulated insulin releases are both considerably elevated (Hales & Kennedy, 1964; Frohman *et al.*, 1969; Steffens, 1969*c*). Lesion-induced hyperinsulinism results from a disturbance of the autonomic control of the endocrine pancreas. Data from several sources agree that the altered secretory response to glucose after a VMN lesion results from a loss of adrenergic inhibitory control of the pancreatic islets.

In rats made diabetic by streptozotocin treatment, transplantation of embryonic islets into the kidney suppressed diabetic symptoms; but the transplanted islets, which can act as functional replacements for the destroyed pancreatic beta-cells, lack innervation. In such rats a VMN lesion does not induce hyperinsulinism, obesity or hyperphagia (Inoue *et al.*, 1978). Sub-diaphragmatic vagotomy, performed sometime after the lesion, blocks and reverses the developing obesity (Powley & Opsahl, 1974). Only complete vagotomy suppresses hyperphagia and obesity; selective vagotomy, sparing the hepatic and coeliac branches, is ineffective (Sawchenko *et al.*, 1981). Thus the gastric field of the vagus nerve is not involved.

Inoue & Bray (1980) and Bray *et al.* (1981) have listed other evidence indicating a reduced sympathetic tone resulting from VMN lesion, such as suppressed basal salivary and gastric secretion. The reduction of the sympathetic activity has been directly confirmed by the demonstration of a reduced turnover of noradrenaline in VMN-lesioned rats (Van der Tuig *et al.*, 1982). Thus, it seems that the VMN lesion introduces an imbalance between the sympathetic and parasympathetic modulation of pancreatic islets, and this leads to hyperinsulinism due to the dominant activity of the vagal facilitation.

Nevertheless, is hyperinsulinaemia a necessary and sufficient cause of obesity? The correlation between insulin level, rate of weight gain, and obesity in individual rats demonstrates that hyperinsulinaemia plays a predominant role (Hustvedt & Løvø, 1972). However, because an impairment of lipolysis is a major contributor to the developing obesity, it is unlikely that the loss of sympathetic inhibition of insulin release is the only factor involved. In intact rats lipolysis induced by fasting, or as a counter-regulatory reaction to fat-store repletion, is brought about by direct sympathetic command of hypoinsulinism and three other adrenergic descending pathways – splanchnic innervation of the liver, adrenergic stimulation of the adrenal medulla, and direct adrenergic innervation of adipose tissue. Thus, a lowered adrenergic tone after the lesion should lead to adiposity, even in the absence of hyperinsulinaemia or when hyperinsulinaemia is prevented. Such is the case in rats made diabetic before the VMN lesion. In streptozotocin- or alloxan-induced diabetes, the symptoms resulting from the lesion are strongly attenuated. Developing adiposity can be clearly observed, however, at least in rats lesioned at weaning (Goldman *et al.*, 1972*a*). Analogous facts have been reported in genetically obese rats and mice (Solomon *et al.*, 1974; Batchelor *et al.*, 1975). A residual 20% hyperphagia in VMN-lesioned rats with transplanted islets is probably an effect of insulin-independent lipogenesis.

The origin of hyperphagia

The dominance of hyperinsulinism over hyperphagia has been fully established. Thus hyperphagia is not a cause but an effect of the neuroendocrine disturbance that determines obesity. Hyperinsulinism and increasing adiposity appear immediately in lesioned weaning rats in the absence of hyperphagia (Frohman & Bernardis, 1968). When lesioned adult rats are pair-fed with unoperated controls, hyperinsulinism and increasing adiposity are observed in lesioned rats, despite the prevention of hyperphagia (Han, 1967). The individual level of hyperinsulinism observed under pair-feeding conditions can be positively correlated with the rate of weight gain when rats are later allowed *ad libitum* feeding. The rate of weight gain can then also be correlated with the level of hyperphagia (Hustvedt & Løvø, 1972). All the data listed in Chapter 6 which show that in the normal rat the nocturnal lipogenesis and the daytime lipolysis are the chief determinants of the associated respective hyperphagia and hypophagia also prove that the enhanced and continuous lipogenesis is the cause of the hyperphagia of hypothalamic lesioned rats. The diversion of ingested metabolites towards synthesis of adipose tissue, as is the case during the dark period in normal rats, permanently accelerates the rate of occurrence of meal initiation by the LH feeding system. Thus the VMN is not involved in the feeding mechanism proper and the LH system and connected structures are not involved in the regulation of body fat mass.

The nature of the VMN liporegulatory mechanism

The VMH is the starting point not only for lipolysis and hypophagia compensating for and repressing an overproduction of body fats but also for the induced extra heat production apparently brought about by the brown adipose tissue (BAT). Under a cafeteria regimen, this extra heat production by the BAT is stimulated by adrenergic innervation originating in the VMN of the hypothalamus (Girardier *et al.*, 1976; Perkins *et al.*, 1981). After VMN lesion, this activation of BAT thermogenesis, during exposure to cold or in limiting the white fat deposition, is impaired (Seydoux *et al.*, 1981). Possibly in this condition of a primary hyperphagia elicited by the cafeteria regimen, the developing obesity elevates the sympathetic tone which normally induces the repressing lipolysis. The lipolysis not manifested in this condition of sensory-induced hyperphagia would normally be relayed by the activation of the BAT. This is also brought about by the obesity-induced increase of sympathetic activity.

Thus, it emerges that hypothalamic VMN are critical sites of the regulation of the body fat mass. It is as if activation of the VMN, in parallel to or determined by some upper weight limit of body fat, corrects lipolysis through descending sympathetic pathways. Inactivation of the VMN, when the lower weight limit of the body fat store is reached, would bring about the converse regulatory pattern. What is the stimulus parallel to, or determined by, the body fat level that could account for such neuronal regulation of the fat reserve? We may disregard the concept of a neural afferent conveying information to the hypothalamus about adipocyte size and number in various depots. Neural afferents from adipose tissues are unknown and apparently do not exist. Rather, the existence of a stimulus acting simultaneously on the central nervous system and on body fats should be explored. Evidence exists that this stimulus might be glucopenia.

VMN glucose-sensitive sites

Evidence has accumulated which suggests that there are neurones, highly responsive to the glucose supply, located within or in the vicinity of the VMN. Furthermore, it has been suggested that glucose responsiveness in VMN neurones or glucose access to these neurones may be insulin dependent.

Jean Mayer and his co-workers (Mayer, 1953) have provided a first line of evidence in their studies of hyperphagia and obesity produced by injection of gold thioglucose (GTG) in mice and by a VMN implant of the same drug in rats (Mayer & Marshall, 1956; Smith & Britt, 1971; Sandrew & Mayer, 1973). Because this neurotoxic agent induces hyperphagia and obesity, treatment must necessarily act on a critical site, the electrolytic destruction of which also produces hyperphagia and obesity. This site is the VMH or its closely connected structures. Three arguments support this assertion. The GTG syndrome of hyperphagia coupled with obesity is similar to the state produced by electrolytic lesion of the VMH. However, the former differs from the latter in various aspects: there is no static phase (Friedman *et al.*, 1963); the diurnal cycle is maintained with high intake at night (Rietveld *et al.*, 1979); and hypophysectomy, which does not prevent the electrolytic-lesion syndrome, prevents the GTG syndrome (Powley & Plocher, 1980).

Although non-hypothalamic sites are lesioned by systemic administration of GTG in mice, particularly the dorsomotor nucleus of the vagus and the nucleus of the tractus solitarius, the main focus of necrosis has been located in the VMH: the deposition of gold in the VMN after systemic administration has been located autoradiographically (Kataoka *et al.*, 1978).

Hyperglycaemia induced by a glucose infusion prior to GTG injection as well as intraventricular phlorizin or 2-DG prevent necrosis, suggesting there is competition for site occupancy or for transmembrane transport Likuski *et al.*, 1967; Debons *et al.*, 1962, 1970). Finally, necrosis, and therefore glucose sensitivity of target neurones or glucose access to them, is insulin dependent. Necrosis does not occur in diabetics or in the presence of insulin antibodies. In diabetics, the lesioning effect is restored after insulin replacement therapy. GTG is also ineffective in fasted hypoinsulinic mice (Debons *et al.*, 1974).

Other techniques have provided results consistent with the GTG action. A gastric load of [^{14}C]glucose is followed by a greater retention of radioactivity and a higher rise in O_2 consumption in the VMH area compared with that in the LH. This effect is reduced in diabetics, is slightly augmented by insulin, and is not modified by previous food deprivation or by the amounts of ^{14}C administered (Panksepp, 1972).

Electrophysiological recordings also confirm the conclusions drawn from the GTG action. The firing rate of single units recorded in the VMN is augmented by an iontophoretic application of glucose. It is again augmented to a greater degree by insulin added to glucose but it is inhibited by insulin alone (Oomura, 1976).

Thus, glucose-sensitive or glucoreceptive sites are present in the VMN and, contrary to those identified in the LH, they appear to be insulin dependent. What is their possible role in the regulation of body fats by the VMN?

Brain insulin receptors

Insulin receptors have been found in the brain. They are segregated in various areas: the olfactory bulb, the arcuate nucleus and the VMN and lateral hypothalamus, the hind-brain in the vicinity of the fourth ventricle (Havrankova *et al.*, 1978, 1981; Oomura & Kita, 1981; Van Houten & Posner, 1981). Some of these receptors, located by autoradiographic techniques, are situated in the wall of capillaries and are therefore on the blood side of the blood–brain barrier. Others are found in circumventricular organs also on this side of the blood–brain barrier. They are situated in axon terminals in the VMN and in cell bodies in the area postrema (Van Houten & Posner, 1981). Both categories bind acutely circulating insulin. Like insulin receptors in other tissues, it is presumed that these brain insulin receptors are down-regulated. As is the case in adipocytes, hepatocytes and blood cells, an elevated plasma insulin concentration induces a reduction in the number and affinity of membrane insulin receptors (down-regulation) and hypoinsulinism gives rise to the opposite trend (up-regulation) (Andreani *et al.*, 1981).

It has been suggested that down- and up-regulation of insulin receptors in the VMH and the effect on the glucose responsiveness of these neurones could account for the role of VMN in the liporegulatory mechanism (Le Magnen, 1981, 1983). Contrary to the action of peripherally injected insulin, the hormone (when infused intraventricularly in baboons and in rats) produces weight loss and hypophagia (Woods *et al.*, 1979; Davis & Brief, 1981). This effect suggests that the weight loss and hypophagia which follow the overweight induced by peripherally injected insulin (PZI treatment or continuous infusion of insulin) result from this central action of insulin developed and masked during the treatment and due to the down-regulation of brain insulin receptors (Le Magnen, 1984). Similarly, it has been suggested that the daytime weight loss and hypophagia within the normal diurnal pattern of rats are effects of the down-regulation of brain insulin receptors by hyperinsulinism during the preceding night. Hyperinsulinism in parallel to its peripheral lipogenic action would induce centrally the regulatory lipolytic pattern by a down-regulation of VMN-insulin receptors. Hypoinsulinism, in parallel to peripherally induced or concurrent lipolysis, would induce the opposite reaction. A down-regulation of insulin receptors may increase the sensitivity of insulin-dependent glucose-sensitive cells to glucopenia. According to this model, these changes in the sensitivity to glucopenia of VMN neurones would be the monitoring signal of the liporegulatory mechanism. Many unexplained phenomena may be interpreted by this model, which, however, has not as yet received direct confirmation (Le Magnen, 1983).

11 Nutrient-specific appetites

The mechanisms described in Chapter 10 ensure an equilibrium between energy intake in the form of foods and energy utilization, i.e. heat production, fat synthesis or other means of energy retention. However, the intake of food, in addition to supplying energy requirements, may provide some specific non-caloric nutrients, such as vitamins, minerals and may also include given amounts of the three macronutrients (carbohydrates, fats and proteins – sources of metabolizable energy), particularly proteins. Such additional nutrient-specific appetites arise as a consequence of the body's need for some elements or molecules which cannot be synthesized biologically. For each of these qualitative needs, there is a quantitative requirement dependent on the particular turnover rate – catabolic loss or retention – of the nutrient and on its synthesis. As the time-course of the equilibrium between loss and intake of energy is affected by the existence of an endogenous energy store and its regulation, the urge to satisfy nutrient-specific needs by intake should be affected by the existence or not, and the capacity, of endogenous reserves of each of the nutrients.

Manifestations of nutrient-specific appetites

Long ago, it was demonstrated that the choice of food in a condition of free selection and response to a mixed diet, as a function of its content, manifested *nutrient-specific appetites* qualitatively and quantitatively adapted, and adaptable, to metabolic requirements. These nutrient-specific responses to types of food are in addition to, or coincidental with, the response based upon their caloric properties. This demonstration came from the extensive studies on rats by Richter (Richter *et al.*, 1937) and later by Scott (Scott, 1948).

Vitamin- and mineral-specific appetites

Given the choice (*a*) between basic macronutrients lacking vitamins and pure solutions of these vitamins or (*b*) between vitamin-free or vitamin-rich diets, rats can exhibit vitamin-specific appetites (Harris *et al.*, 1933; Luria, 1953; Richter & Helfick, 1943; Scott & Quint, 1946; Scott & Verney, 1947; Scott *et al.*, 1950). The amount of, for example, vitamins A, B_1, or B_2, selected is apparently adequate to satisfy the metabolic requirement in the steady-state condition. It increases selectively in rats previously made deficient in the particular vitamin and is seen to be adequate because of the resulting disappearance of the specific symptoms of the deficiency. Thus, vitamin deficiency increases a vitamin-specific appetite as, similarly, food deprivation increases energy intake.

A mineral- or oligo-element-specific appetite can also be exhibited (Richter & Helfick, 1943; Scott *et al.*, 1950; Lewis, 1964; Adam & Dawborn, 1972; Hesse *et al.*, 1979). Sodium-specific appetite, particularly, has been extensively studied in relation to the role of sodium in extracellular and intracellular osmotic pressure and to thirst mechanism. Blood hypertonicity or hypotonicity rapidly induces changes of preference for NaCl solutions as a function of concentration and of the total intake of NaCl (Le Magnen, 1953*b*). The intake of NaCl solution is dramatically increasedd in rats made sodium-deficient by sodium-free diets or abruptly by adrenalectomy (Richter, 1936). An important point is that this increase is more closely related to the salty taste than to the sodium content of the solution. After adrenalectomy and sodium deficiency, rats increase their intake of salty tasting sodium-free salts (Schulkin, 1982). This is identical with the increase of intake of a sweet-tasting saccharin solution under conditions of food deprivation or insulin administration (Le Magnen, 1953*a*).

Macronutrient-specific appetites

Undeprived rats, offered a choice between the three macronutrients, which supply metabolizable energy – carbohydrates, fats and proteins – partition their intake of the three items according to what is adequate for growth and maintenance. However, the question was raised as to whether the response to the three macronutrients, in addition to the response to their respective caloric properties, manifested a carbohydrate-, fat- or protein-specific appetite in relation to a specific metabolic requirement of each of these nutrients. This could be exhibited in the repartitioning of intake in undeprived conditions and/or, more clearly, in conditions of metabolic imbalance creating or exhibiting a specific need for each macronutrient. The response to a mixed diet as a function of its content of the three calorically active materials could also reveal whether the three caloric suppliers were interchangeable or not in various conditions.

Does a carbohydrate-specific appetite exist?

A true learned carbohydrate-specific appetite is questionable (Rozin, 1968). An unlearned specific appetite for sweet-tasting substances does exist. It is not a true and persistent carbohydrate appetite. As just recalled, food deprivation increases saccharin intake. In a choice between two sugar solutions, rats initially prefer the sweetest one; but rapidly, as detailed in previous chapters, these inadequate initial responses are corrected through a conditioning based upon the post-ingestive respective caloric properties. However, under various conditions, rats and other animals – including man – can exhibit selective intake or, conversely, selective rejection of carbohydrate that indicates a regulation of their intake that is dependent on the specific metabolic efficiency of this nutrient. In active compared to non-active rats, and on exposure to cold (Andik *et al.*, 1951; Collier *et al.*, 1969), carbohydrate intake is increased selectively, while after fasting intake of fat and not that of carbohydrate is increased (Andik & Bank, 1954). In diabetic rats which are unable to use carbohydrates as sources of energy, the choice between the three macronutrients results in a fall in intake of carbohydrate and a compensating increase of intake of fat and of protein (Richter & Schmidt, 1941). Diabetic rats given a mixed diet, increase their intake in a manner similar to that of rats given a diet diluted by addition of an inert material (Friedman, 1978; Chapter 3). However, within certain limits, rats can survive and can balance their energy requirement on a carbohydrate-free diet. After such carbohydrate deprivation, the rats do not exhibit an increased intake of carbohydrates, which would indicate a specific need not covered by fats and proteins. On the contrary, after a period of feeding on fat or protein, an induced impairment of carbohydrate metabolism produces a transient decrease in carbohydrate intake, as is the case, for the same reason, after starvation or in diabetes (Lundbaeck & Stevenson, 1947; Randle, 1965; Maji & Ashida, 1978; Blazquez & Quijada, 1968). Again, under certain conditions, fats and carbohydrates are interchangeable as sources of energy. As mentioned in Chapters 5 and 6, in lipolysis the fat supply to the tissues partly relieves glucose utilization and is reflected in the decrease in intake of a carbohydrate-rich diet such as the stock-diet. A pre-load of carbohydrate reduces, within the subsequent meal, fat and protein intake in proportion to the caloric content of the load (Booth, 1974; Geliebter, 1979). A continuous infusion of carbohydrate solution, when supplemented by insulin, may decrease the caloric parenteral supply in proportion to the intake of a complete diet (Rowland *et al.*, 1973).

However, a persistent and unadjusted response to a sweet-tasting carbohydrate solution is sometimes observed. Rats in *ad libitum* conditions, offered their stock-diet and a sucrose solution, persistently increase their intake by 30 to 40% and become obese (Kanarek & Marks-Kaufman, 1979; Hirsch *et al.*, 1982). This, however, cannot be considered as evidence for carbohydrate-specific appetite; rather this is a particular case of the impairment of the attainment of energy balance by intake introduced by a choice of high palatability foods (cafeteria-regimen, see p. 65).

Does a fat-specific appetite exist?

That there is a basal need for fats is not self-evident. Apart from the requirement for essential fatty acids, carbohydrates and proteins can apparently substitute for fats as suppliers of calories. This, however, is not the case when the animal becomes unable to metabolize carbohydrates, permanently (as in diabetes) or transiently (as after starvation), or when fats are preferentially oxidized by the tissues, e.g. in rats during the daytime. A deprivation of fats does not induce a subsequent craving for fats when a choice is presented. However, a puzzling problem is presented by the observation, in some strains of rats and mice, of sustained hyperphagia, and resultant obesity, induced by high-fat diets. A short-lived high responsiveness to fats is analogous to that observed with sweet-tasting solutions. A greasy texture gives rise to an unlearned high palatability response to greasy materials, even when they are inert substances such as vaseline. But, as is the case with saccharin, this response to non-caloric greasy substances is rapidly corrected, which supports the suggestion that a persistent high intake of a high-fat diet is not due to palatability (Hamilton, 1964). After a gastric load of fats, the reduction of intake of the stock-diet during the subsequent hours is less than that induced by a carbohydrate load and the total intake (load plus oral intake) is increased (Geliebter, 1979). This suggests that a part of the caloric intake from fats (which, as it will be seen below, corresponds to the weight gain and fat depot) is not accounted for by mechanisms which maintain the body energy balance in lean tissue.

A protein-specific appetite does exist

There is much data supporting the hypothesis that the intake of proteins is regulated through a mechanism related to some specific metabolic properties of proteins and protein catabolism, and independent of mechanisms by which food intake regulates energy balance (Richter *et al.*, 1937; Scott, 1946; Rozin, 1968).

The main evidence is as follows: In a choice between a low and a high protein diet, e.g. 5% and 40%, respectively, the daily protein intake of rats is remarkably constant relative to the total caloric intake. Changes in the latter may occur by alteration of the choice without inducing a change in the protein intake. It is the case in rats with lesion of the ventromedial nucleus (VMN): they greatly increase their caloric intake and maintain their pre-operative protein intake (Anderson *et al.*, 1979). Under the action of some anorectic drugs, the intake of proteins in a condition of similar choice is increased or decreased independently of changes in the carbohydrate content of the two diets (Wurtman & Wurtman, 1977, 1979). During the diurnal cycle, the pronounced dark–light periodicity of the caloric intake is not associated with an identical periodicity of the protein intake. At the end of the day, meal sizes increase progressively on the two protein-containing diets, while the protein intake remains constant. During the night, the caloric intake from proteins, relative to the total caloric intake of each meal, is lower than for the daytime (Johnson *et al.*, 1979; Leathwood & Arimanana, 1980). Protein deprivation or excessive protein intake during one meal, consisting of either a protein-free or of a high-protein-mixed diet, was followed 45 min later by an increase or a decrease, respectively, in protein intake in a choice during a short meal (Li & Anderson, 1982). Orally deficient or excessive intakes of proteins are approximately compensated by an increase or a decrease in the subsequent 24 h intake (Booth, 1974). Gastric pre-loads or intestinal infusions of amino-acids also reduce the subsequent 2 to 3 h intake of the maintenance diet, as do loads or infusions of the other two macronutrients (Booth, 1972*b*; Novin *et al.*, 1979). The respective roles of caloric and specific properties of proteins in these responses are unclear, as well as the exact nature of the induced changes of the subsequent intake (meal size or meal-to-meal interval changes). The most dramatic observation is that when rats are offered a low-protein, protein-free, or an imbalanced diet, in which only one of the essential amino-acids is lacking, they rapidly refuse to eat this diet. Even after a previous period of food deprivation, this refusal of intake of low protein or imbalanced diet occurs within the first 4 h of presentation (Booth & Simson, 1974). The strong aversion to an amino-acid imbalanced diet is particularly strong. In a choice, rats come to prefer a protein-free to an imbalanced diet (Perez-Zahler & Harper, 1972; Sanahuja & Harper, 1962). A high-protein diet (40% proteins or more) is also aversive, even after fasting, and is rapidly refused by rats, contrary to the results with the high-fat diet (Krauss & Mayer, 1965). A VMN lesion does not prevent such self-starvation.

Mechanisms of nutrient-specific appetites

Vitamins and minerals

Clearly, the depletion of various vitamins and minerals, which is reflected in subnormal plasma concentrations, cannot be active on as many different brain chemosensors controlling specific ingestive responses to them as foods. Except for salty-tasting salts, the substances in foods or in pure solutions are not *per se* discriminative sensory stimuli, that could result in the observed selection and its adjustment to metabolic demands. Thus the old concept of 'partial hunger' of B_1, B_2 vitamin-hungers (Turró, 1914) is entirely irrelevant. The famous experiment of Harris *et al.* (1933), reported in Chapter 4, demonstrates that such specific appetites are dependent on the conditioning of palatability. In this type of conditioning, as with caloric appetites, the conditioned stimulus is generally a sensory cue (the odour or complex flavour of the diet), occasionally or permanently associated with the presence in the diet of the vitamin or mineral. What is then the unconditioned stimulus? It is not the relief of a specific hunger but the recovery of metabolic disturbances induced by the depletion or deficiency. As mentioned earlier, recovery from malaise, specifically recovery from thiamine deficiency, was clearly demonstrated as source of conditioning of a conditioned taste preference (Rozin, 1965; Seward & Greathouse, 1973). Injecting thiamine into a thiamine-deficient rat leads to the development of a preference for a diet lacking the vitamin, when such a diet is paired with the injection (Rozin, 1965; Hasegawa, 1981).

Most vitamin deficiencies manifest themselves in anorexia due to disturbances, e.g. in carbohydrate metabolism, produced by the deficiency and to conditioned taste aversion to all vitamin-free foods as a result of these disturbances (Richter *et al.*, 1938). The establishment and manifestation of the specific appetite is obviously the opposite process.

High-fat diets

The persistent excessive intake of a high-fat diet is presumably due to another mechanism associated with the existence of a fat reserve and of a specific action of long-chain triglycerides. Long-chain free fatty acids hydrolysed from triglycerides during digestion are transported via the lymphatic duct and not via the portal route as are other energy metabolites and medium-chain free fatty acids (MacDonald *et al.*, 1980; Friedman *et al.*, 1983). Presumably as a consequence of these different pathways, pre-loads of medium-chain triglycerides are more satiating than long-chain triglycerides (Geliebter *et al.*, 1983; Gurr *et al.*, 1979; Maggio & Koopmans, 1982). In some strains of rats, at least, long-chain free fatty acids are preferentially diverted into fat depots and are thus unavailable to fuel the lean tissues. The comparison of weight gain of seven strains of rats on high-fat diets showed that the strain-specific weight gain is accounted for by an equivalent strain-specific hyperphagia. Thus a role for a diet-induced thermogenesis in attenuating the induction of obesity, or as a cause of the hyperphagia, was ruled out (Schemmel *et al.*, 1970). It has been proposed that extra-heat production by the brown adipose tissue may explain the lack of obesity observed in rats fed short- and middle-chain triglycerides (Baba *et al.*, 1982). This is unlikely, because there is no evidence that, like long-chain triglycerides, such medium-chain fat diets produce hyperphagia.

Protein intake

The regulation of protein intake involves a more specific and complex mechanism. Like other nutrient-specific appetites, it is clearly a conditioned response to the protein content of the food, based upon associated sensory cues. Rats are given alternately two forms of a protein-free diet with different odours added. The consumption of one of them is preceded by a gastric load of a mixture of essential amino-acids, and the other one by a gastric load of saline. Within the first 2 h of the first presentation, an aversion was manifested for the latter form while a preference was developed for the former. In a final choice, not associated with gastric loads, rats preferred the protein-free diet labelled by the odour previously associated with the amino-acid gastric load. In a similar experiment, rats were also offered two forms of the diet with odours added but each lacking one of the essential amino-acids. A preference was developed for the form associated with a gastric loading of the missing amino-acid. The other form, followed by gastric loading of saline, was readily rejected (Simson & Booth, 1973*a*, *b*, 1974*a*, *b*).

What is the post-ingestive sensing mechanism (acting either as a result of an unconditioned stimulus or in satiation of intake of a specific protein) which can explain such rapid responses to low protein or amino-acid imbalanced diets? The satiation process is suggested by the remarkable finding in cats of vagal fibres responding to amino-acids infused into the duodenum, some of them responding only to one or two amino-acids (Jeanningros, 1982). The same infusion alters the firing rate of single neurones in the lateral hypothalamus. So these vagal fibres may convey the information on the amino-acid content of the intestine to the lateral hypothalamic area (Jeanningros, 1983). However, the role played by this 'intestinal taste' does not exclude a part in the systemic depletion and repletion of proteins. The intake of protein-free, low- or high-protein or amino-acid-imbalanced diet is rapidly followed by changes in the plasma amino-acid spectrum, and this, in turn, is readily reflected by the concentrations of amino-acids in the brain (Wurtman & Wurtman, 1977). A negative correlation has been found between the ratio of tryptophan to neutral amino-acids in the blood and the protein intake (Ashley & Anderson, 1975). A high-protein diet, contrary to a high-fat diet, in the absence of possible storage, is rapidly toxic and induces the conditioned taste aversion. A low protein diet strongly impairs insulin release and glucose utilization (Beaton *et al.*, 1964, 1968; Royle *et al.*, 1982; Usami *et al.*, 1982). This can also explain the rapidly acquired aversion for the low-protein mixed diet.

Specific brain mechanisms responsible for the maintenance of the protein balance and its regulation by selective protein intake have been investigated. As mentioned above, the hypothalamic VMN, which are involved in the regulation of body fat mass, cannot be involved in protein intake since the protein intake is maintained after lesion (Anderson *et al.*, 1979). Lesions of various structures have failed to modify the regulation of protein intake (Leung & Rogers, 1969, 1970, 1971, 1973). Only the lesion of the amygdala, which is involved in the conditioning of taste preferences and aversions, impairs the aversion to an amino-acid imbalanced diet (Meliza *et al.*, 1981; Kolakowska *et al.*, 1984). Finally, it has been shown that amino-acids iontophoretically applied, altered the firing rate of single neurones in the lateral hypothalamus (Wayner *et al.*, 1975; Jeanningros, 1984). It was suggested that such responses could be evidence for local chemosensors of aminoacidaemia. However, lesion of the area does not induce any deficiency in the regulation of protein intake in rats recovered from aphagia (Wayner *et al.*, 1979).

Serotonergic structures have been implicated (Ashley *et al.*, 1979; Wurtman & Wurtman, 1979). However, the inter-individual differences in the protein intake of rats was not found to be correlated to the brain serotonin content (Leathwood & Arimanana, 1980). It is assumed that the diffuse dysfunction of the brain due to the lack of amino-acids needed for the synthesis of enzymes and of all neurotransmitters may be involved as a positive or negative reinforcer of the protein oral intake.

Conclusion

The limited length of this book has not permitted me to review all the various aspects of hunger and food intake and homeostatic mechanisms by which body nutrition is achieved through behaviour: some important issues such as drinking–feeding relationships, for example, have been dropped. Neurochemical approaches to brain functions could not be briefly analysed because the many contradictory results make it difficult to draw a clear picture. More generally many elegant studies should have been cited in addition to those referred to in the book. Some experiments have been selected as examples among many which gave identical results. My guideline was to extract a comprehensive overview of the present knowledge from an enormous literature on an interdisciplinary topic dealing simultaneously with neurophysiology, physiological psychology, endocrinology and nutrition.

Thanks to striking advances during the last decades, this knowledge permits a new insight by integrating the various approaches. Some points seem to have been definitely established and are no longer questionable. Many of them are conceptual and methodological in nature. This is so, for example, in the separation of distinct mechanisms governing respectively meal sizes and the initiation of feeding; the systemic stimulus of hunger arousal and of palatability of foods as combined stimuli to eat; and the regulations of body fat mass and of food intake, and their interactions. Much previous and persisting confusion has resulted from the ignorance of such basic concepts and from a vague terminology, various phenomena being covered by the same word, and identical ones by different words.

From this solid background of well-separated phenomena and of clearly identified mechanisms, some points (at present suggested by convincing evidence) will still have to be confirmed. Many aspects of brain mechanisms remain obscure and need further studies dealing with, for instance, the role of the hindbrain and of glucose-sensitive sites around the fourth ventricle compared to the role of other glucose-sensitive sites identified in the hypothalamus and that of brain insulin receptors.

Similarly, brain mechanisms involved in the learning of the differential palatability of foods, which is almost entirely unknown, should be a decisive field of future investigations. These investigations, at present limited to 'conditioned taste aversion', should lead to an understanding of the main characteristics of all behaviours and of brain functions, i.e. 'anticipations' of behavioural responses appropriate to survival from a sensory analysis of the environment and as a result of previous experience.

This essential brain mechanism leads to a 'wisdom of the body' through subcortical neuronal networks, which is exhibited in feeding behaviour by a free selection and amount of food eaten adjusted to and reflecting bodily needs. These mechanisms have been fixed in genetic programmes through natural selection. In higher vertebrates, a development of the neocortex adds a wisdom derived from individual experience and cognitive processes.

The cumulated experience of humanity and the social transmission of knowledge are added in man to the individual experience. Unfortunately, in feeding behaviour as in other instances, this cognitive guidance of behaviour added to or substituted for the 'wisdom of the body' is not always (at least at a given state of knowledge) necessarily beneficial. Hunger in the world is still a tragic problem; yet in developed countries men become obese.

References

Adam, W. R. & Dawborn, J. K. (1972). Effect of potassium depletion on mineral appetite in the rat. *J. Comp. Physiol. Psychol.*, **78**, 51–8.

Ahlskog, J. E., Randall, P. K. & Hoebel, B. G. (1975). Hypothalamic hyperphagia: dissociation from hyperphagia following destruction of noradrenergic neurons. *Science*, **190**, 399–400.

Aleksanyan, Z. A., Buresova, O. & Bures, J. (1976). Modification of unit responses to gustatory stimuli by conditioned taste aversion in rats. *Physiol. Behav.*, **17**, 173–9.

Allison, J. & Castellan, N. J. (1970). Temporal characteristics of nutritive drinking in rats and humans. *J. Comp. Physiol. Psychol.*, **70**, 116–25.

Almli, C. R. (1978). The ontogeny of feeding and drinking effects of early brain damage. *Neurosci. Biobehav. Rev.*, **2**, 281–300.

Anand, B. K. & Brobeck, J. R. (1951). Localisation of a 'feeding center' in the hypothalamus of the rat. *Proc. Soc. Exp. Biol. Med.*, **77**, 323–4.

Anand, B. K., Chhina, G. S. & Singh, B. (1962). Effect of glucose on the activity of hypothalamic 'feeding centers'. *Science*, **138**, 597–8.

Anand, B. K., Dua, S. & Shoeberg, K. (1955). Hypothalamic control of food intake in cats and monkeys. *J. Physiol.* (*Lond.*), **127**, 143–53.

Anderson, G. H., Leprohon, C., Chambers, J. W. & Coscina, D. V. (1979). Intact regulation of protein intake during the development of hypothalamic or genetic obesity in rats. *Physiol. Behav.*, **23**, 751–6.

Andik, I. & Bank, J. (1954). Influence de la température sur la prise alimentaire du rat normal et du rat traité à la thyroxine. *Acta Physiol. Acad. Sci. Hung.*, **6** (suppl.), 37.

Andik, I., Donhoffer, S. Z., Moring, I. & Szentes, J. (1951). The effect of starvation on food intake and selection. *Acta Physiol. Acad. Sci. Hung.*, **2**, 363–8.

Andreani, D., De Pirro, R., Lauro, R., Olefsky, J. M. & Roth, J. (eds.) (1981). *Current Views and Insulin Receptors.* Academic Press: New York.

Anika, S. M., Houpt, T. R. & Houpt, K. A. (1980). Insulin as a satiety hormone. *Physiol. Behav.*, **25**, 21–4.

Anliker, J. & Mayer, J. (1956). Operant technique for studying feeding–fasting patterns in normal and obese mice. *J. Applied Physiol.*, **8**, 667–70.

Antin, J., Gibbs, J. & Smith, G. P. (1977). Intestinal satiety requires pregastric food stimulation. *Physiol. Behav.*, **18**, 421–6.

Aparicio, N. J., Puchulu, F. E., Gagliardino, J. J., Ruiz, M. D., Llorens, J. M., Ruiz, J., Lamas, A. & De Miguel, R. (1974). Circadian variation of the blood glucose, plasma insulin and human growth hormone levels in response to an oral glucose load in normal subjects. *Diabetes*, **23**, 132–7.

Apfelbaum, M., Bostsarron, J., Lacatis, D. & Duret, F. (1972*a*). Augmentation de la consommation d'oxygène chez 8 sujets bien portants, recevant pendant 15 jours un supplément de 1.500 calories. *J. Physiol.* (*Paris*), **65**, 91A.

Apfelbaum, M., Reinberg, A., Assan, R. & Lacatis, D. (1972*b*). Persistance du rythme circadien de l'insuline du glucagon et cortisol plasmatique ainsi que la consommation d'O_2 et du quotient respiratoire pendant une restriction calorique par diète protéique. *Ann. Endocr.*, **33**, 274.

Applegate, E. A., Upton, D. E. & Stern, J. S. (1982). Food intake, body composition and blood lipids following treadmill exercise in male and female rats. *Physiol. Behav.*, **28**, 917–20.

Aravich, P. F. & Sclafani, A. (1980). Dietary preference in rats fed bitter tasting quinine and sucrose octa-acetate adulterated diets. *Physiol. Behav.*, **25**, 157–60.

(1983). Paraventricular hypothalamic lesions and medial hypothalamic knife cuts produce similar hyperphagia syndromes. *Behav. Neurosci.*, **97**, 970–83.

Ardisson, J. L., Dolisi, C., Ozon, C. & Crenesse, D. (1981). Caractéristiques des prises d'eau et d'aliments spontanées chez des chiens en situation *ad lib. Physiol. Behav.*, **26**, 361–71.

Armitage, G., Hervey, G. R., Rolls, B. J., Rowe, E. A. & Tobin, G. (1981). Energy balance in 'cafeteria'-fed adult rats. *J. Physiol.* (*Lond.*), **316**, 48P–49P.

Armitage, G., Hervey, G. R. & Tobin, G. (1979). Energy expenditure of rats tube-fed at different energy levels. *J. Physiol.* (*Lond.*), **290**, 17P–18P.

Ashe, J. & Nachman, M. (1980). Neural mechanisms in taste aversion learning. In *Progress in Psychobiology & Physiological Psychology*, ed. J. Sprague & A. Epstein, vol. 9, pp. 233–62. Academic Press: New York.

Ashley, D. V. M., Coscina, D. V. & Anderson, G. H. (1979). Selective decrease in protein intake following brain serotonin depletion. *Life Sci.*, **24**, 973–84.

Ashley, D. V. & Anderson, G. H. (1975). Food intake regulation in the weanling rat: effects of the most limiting essential amino-acids of gluten, casein and zein on the self selection of protein and energy. *J. Nutr.*, **105**, 1405–11.

Atrens, D. M., Sinden, J. D., Penicaud, L., Devos, M. & Le Magnen, J. (1985). Hypothalamic hypermetabolism: a possible mechanism for diet-induced thermogenesis. *Physiol. Behav.*, in press.

Auffray, P. & Marcilloux, J.-C. (1983). Etude de la séquence alimentaire du porc adulte. *Reprod. Nutr. Develop.*, **23**, 517–24.

Baba, N., Bracco, E. F. & Hashim, S. A. (1982). Enhanced thermogenesis and diminished deposition of fat in response to overfeeding with diet containing medium chain triglyceride. *Am. J. clin. Nutr.*, **35**, 678–82.

Baillie, P. (1977). Patterns of ingestion of rats adjusting to a controlled-feeding schedule of 4 hr per day. *J. Physiol.* (*Lond.*), **273**, 35P.

Baillie, P. & Morrison, S. D. (1963). The nature of the suppression of food intake by lateral hypothalamic lesions in rats. *J. Physiol.* (*Lond.*), **165**, 227–45.

Balagura, S. (1968). Conditioned glycemic responses in the control of food intake. *J. Comp. Physiol. Psychol.*, **65**, 30–2.

Balagura, S. & Devenport, L. D. (1970). Feeding patterns of normal and ventromedial hypothalamic lesioned male and female rats. *J. Comp. Physiol. Psychol.*, **71**, 357–64.

Balagura, S., Harrell, L. E. & Roy, E. (1975). Effect of the light–dark cycle on neuroendocrine and behavioral responses to schedule feeding. *Physiol. Behav.*, **15**, 245–8.

Balagura, S. & Hoebel, B. G. (1967). Self-stimulation of the lateral hypothalamus modified by insulin and glucagon. *Physiol. Behav.*, **2**, 337–40.

Balagura, S. & Kanner, M. (1971). Hypothalamic sensitivity to 2-deoxy-D-glucose and glucose: effects on feeding behavior. *Physiol. Behav.*, **7**, 251–6.

Balagura, S., Kanner, M. & Harrell, L. E. (1975). Modification of feeding patterns by glucodynamic hormones. *Behav. Biol.*, **13**, 457–66.

Balasse, E. O. & Neef, M. A. (1974). Operation of the 'glucose-fatty acid cycle' during experimental elevations of plasma free fatty acid levels in man. *Eur. J. Clin. Invest.*, **4**, 247–52.
Bare, J. K. & Cicalla, G. (1960). Deprivation and time of testing as determinants of food intake. *J. Comp. Physiol. Psychol.*, **53**, 151–4.
Barr, H. G. & McCracken, K. J. (1982). Absence of 'diet-induced thermogenesis' in growing rats kept at 20° and offered a varied diet. *Proc. Nutr. Soc.*, **41**, 63A.
Batchelor, B. R., Stern, J. S., Johnson, P. R. & Mahler, R. J. (1975). Effects of streptozotocin on glucose metabolism insulin response and adiposity in *ob/ob* mice. *Metabolism*, **24**, 77–92.
Beaton, J. R., Feleki, V. & Stevenson, J. A. F. (1968). Factors in reduced food intake of rats fed a low-protein diet. *Can. J. Physiol. Pharmac.*, **46**, 19–24.
Beaton, J. R., Feleki, V., Szlavko, A. J. & Stevenson, J. A. F. (1964). Meal-eating and lipogenesis *in vitro* in rats fed a low-protein diet. *Can. J. Physiol. Pharmac.*, **42**, 665–70.
Bellinger, L. L. & Mendel, V. E. (1975). Effect of deprivation and time of refeeding on food intake. *Physiol. Behav.*, **14**, 43–6.
Bellinger, L. L. & Williams, F. E. (1981). The effects of liver denervation on food and water intake in the rat. *Physiol. Behav.*, **26**, 663–73.
Bellisle, F. & Le Magnen, J. (1980). The analysis of human feeding patterns: the edogram. *Appetite*, **1**, 141–50.
Bellisle, F., Louis-Sylvestre, J., Demozay, F., Blazy, D. & Le Magnen, J. (1983). Reflex insulin response associated to food intake in human subjects. *Physiol. Behav.*, **31**, 515–22.
Bellisle, F., Lucas, F., Amrani, R. & Le Magnen, J. (1984). Deprivation, palatability and the micro-structure of meals in human subjects. *Appetite*, **5**, 85–94.
Berger, P. & Le Magnen, J. (1957). Etude de la faim et des appétits chez l'homme placé dans des conditions extrêmes et prolongées de dénutrition. *C.R. Acad. Sci. (Paris)*, **244**, 494–6.
(1960). La faim et les appétits chez l'homme en état de semi-inanition. *Ann. Nutrit. Alim.*, **14**, 101–33.
Bernstein, I. L. (1976*a*). Spontaneous activity and meal patterns. In *Hunger: Basic Mechanisms and Clinical Implications*, ed. D. Novin, W. Wyrwicka & G. A. Bray, pp. 219–24. Raven Press: New York.
(1976*b*). Ontogeny of meal pattern of the rats. *J. Comp. Physiol. Psychol.*, **90**, 1126–32.
(1981). Meal patterns in 'free running humans'. *Physiol. Behav.*, **27**, 621–4.
Bernstein, I. L. & Goehler, L. E. (1983). Vagotomy produces learned food aversions in the rat. *Behav. Neurosci.*, **97**, 585–94.
Bernstein, I. L. & Vitiello, M. V. (1978). The small intestine and the control of meal patterns of the rat. *Physiol. Behav.*, **20**, 417–22.
Bernstein, I. L. & Woods, S. C. (1980). Ontogeny of cephalic insulin release by the rat. *Physiol. Behav.*, **24**, 529–32.
Bernstein, L. M. & Grossman, M. I. (1956). An experimental test of the glucostatic theory of regulation of food intake. *J. Clin. Invest.*, **35**, 627–33.
Berridge, K., Grill, H. J. & Norgren, R. (1981). Relation of consummatory response and preabsorptive insulin release to palatability and learned taste aversions. *J. Comp. Physiol. Psychol.*, **95**, 363–82.
Berthoud, H. R. & Mogenson, G. J. (1977). Ingestive behavior after intracerebral and

intracerebroventricular infusions of glucose and 2-deoxy-D-glucose. *Am. J. Physiol.*, **233** (*Regulatory Integrative comp. Physiol.*, **2**), R127–R133.

Bessard, T., Schutz, Y. & Jequier, E. (1982). Reduced dietary induced thermogenesis in obese women. *Int. J. Vitam. Nutr. Res.*, **52**, 210–11.

Bestley, J. W., Bramley, P. N., Dobson, P. M. S., Mahanty, A. & Tobin, G. (1982). Energy balance in 'cafeteria'-fed young 'Charles River' Sprague Dawley rats. *J. Physiol. (Lond.)*, **330**, 70P–71P.

Birch, L. L. & Marlin, D. W. (1982). I don't like it; I never tried it: effects of exposure on two-year-old children's food preferences. *Appetite*, **3**, 353–60.

Blass, E. M. & Teicher, M. H. (1980). Suckling. *Science*, **210**, 15–21.

Blass, E. M., Teicher, M. H., Cramer, C. P., Bruno, J. P. & Hall, W. G. (1977). Olfactory, thermal, and tactile controls of suckling in preauditory and previsual rats. *J. Comp. Physiol. Psychol.*, **91**, 1248–60.

Blaza, S. & Garrow, J. S. (1983). Thermogenic response to temperature exercise and food stimuli in lean and obese women studied by 24 hr direct calorimetry. *Br. J. Nutr.*, **49**, 171–80.

Blazquez, E. & Quijada, C. L. (1968). The effect of a high-fat diet on glucose, insulin sensitivity and plasma insulin in rats. *J. Endocrinol.*, **42**, 489–94.

Blundell, J. E. & Herberg, L. J. (1968). Relative effects of nutritional deficit and deprivation period on rate of electrical self-stimulation of lateral hypothalamus. *Nature*, **219**, 627–8.

Bodnar, R. J., Merrigan, K. P. & Wallace, M. M. (1981). Analgesia following intraventricular administration of 2-deoxy-D-glucose. *Pharmac. Biochem. Behav.*, **14**, 579–82.

Bolles, R. C. (1962). The readiness to eat and drink: the effect of the deprivation conditions. *J. Comp. Physiol. Psychol.*, **55**, 230–4.

Booth, D. A. (1968). Effects of intrahypothalamic glucose injection on eating and drinking elicited by insulin. *J. Comp. Physiol. Psychol.*, **65**, 13–16.

(1972*a*). Satiety and behavioral caloric compensation following intragastric glucose loads in the rat. *J. Comp. Physiol. Psychol.*, **78**, 412–32.

(1972*b*). Postabsorptively induced suppression of appetite and the energostatic control of feeding. *Physiol. Behav.*, **9**, 199–202.

(1972*c*). Conditioned satiety in rats. *J. Comp. Physiol. Psychol.*, **81**, 457–71.

(1974). Food intake compensation for increase or decrease of the protein content of the diet. *Behav. Biol.*, **12**, 31–40.

(1977). Appetite and satiety as metabolic expectancies. In *Food Intake and Chemical Senses*, ed. Y. Katsuki, M. Sato, S. F. Takagi & Y. Oomura, pp. 317–31. University of Tokyo Press: Tokyo.

(1979). Hunger and satiety as conditioned reflexes. *Arch. Neurol.*, **36**, 866.

Booth, D. A. & Campbell, C. S. (1975). Relation of fatty acids to feeding behaviour: effects of palmitic acid infusions, lighting variation and pent-4-enoate, insulin or propranolol injection. *Physiol. Behav.*, **15**, 523–36.

Booth, D. A., Coons, E. E. & Miller, N. E. (1969). Blood glucose responses to electrical stimulation of the hypothalamic feeding area. *Physiol. Behav.*, **4**, 991–1002.

Booth, D. A. & Davis, J. D. (1973). Gastrointestinal factors in the acquisition of oral sensory control of satiation. *Physiol. Behav.*, **11**, 23–30.

Booth, D. A. & Jarman, S. P. (1975). Ontogeny and insulin-dependence which follows carbohydrate absorption in the rat. *Behav. Biol.*, **15**, 159–72.

Booth, D. A., Lee, M. & MacAleavey, C. (1976). Acquired sensory control of satiation in man. *Br. J. Psychol.*, **67**, 137–47.
Booth, D. A., Mather, P. & Fuller, J. (1982). Starch content of associatively conditioned human appetite and satiation, indexed by intake and eating pleasantness of starch-paired flavours. *Appetite*, **3**, 163–84.
Booth, D. A. & Pitt, M. E. (1968). The role of glucose in insulin-induced feeding and drinking. *Physiol. Behav.*, **3**, 447–53.
Booth, D. A. & Simson, P. C. (1974). Taste aversion induced by an histidine-free amino-acid load. *Physiol. Psychol.*, **2** (3A), 349–51.
Boyle, P. C., Storlien, L. H. & Keesey, R. E. (1978). Increased efficiency of food utilization following weight loss. *Physiol. Behav.*, **21**, 261–4.
Bray, G. A., Inoue, S. & Nishizawa, Y. (1981). Hypothalamic obesity: the autonomic hypothesis and the lateral hypothalamus. *Diabetologia*, **20** (Suppl.), 366–78.
Brobeck, J. R. (1946). Mechanism of the development of obesity in animals with hypothalamic lesions. *Physiol. Rev.*, **26**, 541–59.
(1948). Food intake as a mechanism of temperature regulation. *Yale J. Biol. Med.*, **20**, 545–52.
(1955). Neural regulation of food intake. *Ann. N.Y. Acad. Sci.*, **63**, art. 1, 44–55.
Brooks, S. L., Rothwell, N. J. & Stock, M. J. (1981). Increased resistance to obesity following early exposure to the cafeteria diet. *Proc. Nutr. Soc.*, **40**, A58.
Cabanac, M. (1979). Sensory pleasure. *Quart. Rev. Biol.*, **54**, 29P.
Campbell, R. G., Hashim, S. A. & Van Ittalie, T. B. (1971). Studies of food-intake regulation in man: responses to variations in nutritive density in lean and obese subjects. *New England J. Med.*, **285**, 1402–6.
Campbell, S. & Davis, J. D. (1974). Peripheral control of food intake: interaction between test diet and postingestive chemoreception. *Physiol. Behav.*, **12**, 377–84.
Campfield, L. A. & Smith, F. J. (1983). Neural control of insulin secretion: interaction of norepinephrine and acetylcholine. *Am. J. Physiol.*, **244**, R629–R634.
Campfield, L. A., Smith, F. J. & Le Magnen, J. (1983). Altered endocrine pancreatic function following vagotomy: possible behavioral and metabolic bases for assessing completeness of vagotomy. In *Vagal Nerve Function: Behavioral and Methodological Considerations*, ed. J. G. Kral, T. L. Powley & C. McC. Brooks, pp. 283–300. Elsevier Science Pub.: Amsterdam.
Campfield, L. A., Smith, F. J. & Brandon, P. (1984). Partial blockade of pre-meal decline in blood glucose delays feeding. In *The Neural and Metabolic Bases of Feeding*, Proceedings of a Meeting of the Neurosciences Society, California.
Cannon, W. B. & Washburn, A. L. (1912). An explanation of hunger. *Am. J. Physiol.*, **29**, 441–54.
Capretta, P. J. (1964). Saccharin consumption and the reinforcement issue. *J. Comp. Physiol. Psychol.*, **57**, 448–50.
Carpenter, R. G. & Grossman, S. P. (1983*a*). Early streptozotocin diabetes and hunger. *Physiol. Behav.*, **31**, 175–8.
(1983*b*). Reversible obesity and plasma fat metabolites. *Physiol. Behav.*, **30**, 51–6.
(1983*c*). Plasma fat metabolites and hunger. *Physiol. Behav.*, **30**, 57–63.
Chaput, M. & Holley, A. (1976). Olfactory bulb responsiveness to food odour during stomach distension in the rat. *Chem. Senses Flavor*, **2**, 189–201.
Cheng, M. F., Rozin, P. & Teitelbaum, P. (1971). Starvation retards development of food and water regulations. *J. Comp. Physiol. Psychol.*, **76**, 206.

Christie, M. J. & Chesher, G. B. (1982). Physical dependence on physiologically released endogenous opiates. *Life Sci.*, **30**, 1173–7.

Cohn, D. & Joseph, D. (1962). Influence of body weight and body fat on appetite of 'normal', lean and obese rats. *Yale J. Biol.*, **34**, 598–607.

Coll, M., Meyer, A. & Stunkard, J. (1979). Obesity and food choices in Public places. *Arch. Gen. Psychol.*, **36**, 795–7.

Collier, G., Leshner, A. I. & Squibb, R. L. (1969). Dietary self-selection in active and non-active rats. *Physiol. Behav.*, **4**, 79–82.

Coons, E. E. & Cruce, J. A. F. (1968). Lateral hypothalamus: food current intensity and maintaining self-stimulation of hunger. *Science*, **159**, 1117–19.

Cooper, S. J. (1980). Naloxone: effects on food and water consumption in the non-deprived and deprived rat. *Psychopharmacology*, **71**, 1–6.

Cottle, W. & Carlson, L. D. (1954). Adaptive changes in rats exposed to cold. Caloric exchange. *Am. J. Physiol.*, **178**, 305.

Cramer, C. P. & Blass, E. M. (1983). Mechanisms of control of milk intake in suckling rats. *Am. J. Physiol.*, **245**, R154–R160.

Curry, D. L. & Joy, R. M. (1974). Direct CNS modulation of insulin secretion. *Endocr. Res. Commun.*, **1**, 229–38.

Danguir, J. & Nicolaïdis, S. (1979). Action des perfusions alternées d'insuline et d'adrénaline. *J. Physiol.* (*Paris*), **75**, 1A.

Dauncey, M. J. (1980). Metabolic effects of altering the 24 hr energy intake in man, using direct and indirect calorimetry. *Br. J. Nutr.*, **43**, 257–70.

Davies, R. F. (1977). Long–short term regulation of feeding patterns in the rat. *J. Comp. Physiol. Psychol.*, **91**, 574–85.

Davies, R., Nakajima, S. & White, N. (1974). Enhancement of feeding produced by stimulation of the ventromedial hypothalamus. *J. Comp. Physiol. Psychol.*, **86**, 414–19.

Davis, J. D. & Brief, D. J. (1981). Chronic intraventricular insulin infusion reduces food intake and body weight in rats. *Abst. Soc. Neurosci.*, **7**, 655.

Davis, J. D. & Campbell, C. S. (1973). Peripheral control of meal size in the rat: effect of sham feeding in meal size and drinking rate. *J. Comp. Physiol. Psychol.*, **83**, 379–87.

Davis, J. D., Collins, B. J. & Levine, M. W. (1976). Peripheral control of meal size: interaction of gustatory stimulation and postingestional feedback. In *Hunger: Basic Mechanisms and Clinical Implication*, ed. D. Novin, W. Wyrwicka & G. A. Bray, pp. 395–408. Raven Press: New York.

Davis, J. D., Wirtschafter, D., Asin, K. E. & Brief, D. (1981). Sustained intracerebroventricular infusion of brain fuels reduced body weight and food intake in rats. *Science*, **212**, 81–3.

Debons, A. F., Krimsky, I., From, A. & Cloutier, R. J. (1970). Site of action of gold thioglucose in the hypothalamic satiety center. *Am. J. Physiol.*, **219**, 1397–402.

Debons, A. F., Krimsky, I., From, A. & Pattinian, H. (1974). Phlorizin inhibition of hypothalamic necrosis induced by gold thioglucose. *Am. J. Physiol.*, **226**, 574–8.

Debons, A. F., Silver, L., Cronkite, E. P., Johnson, H. A., Brecher, G., Tenzer, D. & Schwartz, I. L. (1962). Localization of gold in mouse brain in relation to gold thioglucose obesity. *Am. J. Physiol.*, **202**, 743–50.

De Castro, J. M. (1975). Meal pattern correlations: facts and artifacts. *Physiol. Behav.*, **15**, 13–17.

De Castro, J. M. & Balagura, S. (1976). A preprandial intake pattern in weanling rats ingesting a high fat diet. *Physiol. Behav.*, **17**, 404–5.

Delgado, J. M. R. & Anand, B. K. (1953). Increase of food intake induced by electrical stimulation of the lateral hypothalamus. *Am. J. Physiol.*, **172**, 162–8.
Desor, J. A., Maller, O. & Turner, R. (1973). Taste in acceptance of sugars by human infants. *J. Comp. Physiol. Psychol.*, **84**, 496–501.
Deutsch, J. A. & Gonzalez, M. F. (1980). Gastric nutrient content signals satiety. *Behav. Neural. Biol.*, **30**, 113–16.
Deutsch, J. A., Gonzalez, M. F. & Young, M. G. (1980). Two factors control meal size. *Brain Res. Bull.*, **5** (Suppl. 4), 55–8.
Deutsch, J. A. & Hardy, W. T. (1977). Cholecystokinin produces bait shyness in rats. *Nature*, **266**, 196.
Deutsch, J. A., Thiel, T. R. & Greenberg, L. H. (1978). Duodenal motility after cholecystokinin injection or satiety. *Behav. Biol.*, **24**, 393–9.
Deutsch, J. A. & Wang, M. J. (1977). The stomach as a site for rapid nutrient reinforcement sensors. *Science*, **195**, 89–90.
Deutsch, R. (1974). Conditioned hypoglycemia: a mechanism for saccharin-induced sensitivity to insulin in the rat. *J. Comp. Physiol. Psychol.*, **86**, 350–8.
Devos, M. (1981). La périodicité alimentaire du rat blanc: sa signification physiologique. Doctorat d'Université – Paris.
Dum, J., Gramsch, C. H. & Herz, A. (1983). Activation of hypothalamic beta endorphin pools by reward induced by highly palatable food. *Pharmacol. Biochem. Behav.*, **18**, 443–7.
Duncan, I. J. H., Horne, A. R., Hughes, B. O. & Wood-Gush, D. G. M. (1970). The pattern of food intake in female brown leghorn fowls as recorded in a Skinner box. *Anim. Behav.*, **18**, 245–55.
Durnin, J. V. G. A. (1957). The day to day variation in individual food intake and energy expenditure. *J. Physiol. (Lond.)*, **136**, 34P.
(1961). 'Appetite' and the relatioinship between expenditure and intake of calories in man. *J. Physiol. (Lond.)*, **156**, 294–306.
Durrer, J. L. & Hannon, J. P. (1962). Seasonal variations in caloric intake of dogs living in an arctic environment. *Am. J. Physiol.*, **202**, 375–8.
Epstein, A. N. & Teitelbaum, P. (1962). Regulation of food intake in the absence of taste, smell and other oro-pharyngeal sensations. *J. Comp. Physiol. Psychol.*, **55**, 753–9.
(1967). Specific loss of the hypoglycemic control of feeding in recovered lateral rats. *Am. J. Physiol.*, **213**, 1159–67.
Esposito, R. U. & Kornetsky, C. (1978). Opioids and rewarding brain stimulation. *Neurosci. Biobehav. Rev.*, **2**, 115–22.
Ettenberg, A., Sgro, S. & White, N. (1982). Algebraic summation of the affective properties of a rewarding and an aversive stimulus in the rat. *Physiol. Behav.*, **28**, 873–7.
Ettenberg, A. & White, N. (1978). Conditioned taste preferences in the rat induced by self-stimulation. *Physiol. Behav.*, **21**, 363–8.
Ettinger, R. H. & Staddon, J. E. R. (1982). Decreased feeding associated with acute hypoxia in rats. *Physiol. Behav.*, **29**, 455–8.
Evans, H. L. (1971). Rats activity: influence of light–dark cycle, food presentation and deprivation. *Physiol. Behav.*, **7**, 455–60.
Faust, I. M., Johnson, P. R. & Hirsch, J. (1977). Adipose tissue regeneration following lipectomy. *Science*, **197**, 391–2.
Finger, F. W. (1951). The effect of food deprivation and subsequent satiation upon general activity in the rat. *J. Comp. Physiol. Psychol.*, **44**, 557–64.

Fischer, U., Hommel, H., Ziegler, M. & Jutzi, E. (1972). The mechanism of insulin secretion after oral glucose administration: III. Investigations on the mechanism of a reflectory insulin mobilization after oral stimulation. *Diabetologia*, **8**, 385–90.
Fleming, A. S. (1976*a*). Control of food intake in the lactating rat: role of suckling and hormones. *Physiol. Behav.*, **17**, 841–8.
(1976*b*). Ovarian influences on food intake and body weight regulation in lactating rats. *Physiol. Behav.*, **17**, 969–77.
Fonberg, E. & Sychowa, B. (1968). Effect of partial lesions of the amygdala in dogs. I. Aphagia. *Acta Biol. Exp.*, **28**, 35–46.
Friedman, G., Waye, J. D. & Janowitz, H. D. (1963). Persistent dynamic phase of aurothioglucose obesity. *Am. J. Physiol.*, **205**, 919–21.
Friedman, M. (1978). Hyperphagia in rats with experimental diabetes mellitus: a response to a decreased supply of utilizable fuels. *J. Comp. Physiol. Psychol.*, **92**, 109–17.
Friedman, M. I., Edens, N. K. & Ramirez, I. (1983). Differential effects of medium- and long-chain triglycerides on food intake of normal and diabetic rats. *Physiol. Behav.*, **31**, 851–6.
Frohman, L. A. & Bernardis, L. L. (1968). Growth hormone and insulin levels in weanling rats with ventromedial hypothalamic lesions. *Endocrinology*, **82**, 1125–32.
(1971). Effect of hypothalamic stimulation on plasma glucose, insulin, and glucagon levels. *Am. J. Physiol.*, **221**, 1596–1603.
Frohman, L. A., Bernardis, L. L., Schnatz, J. D. & Burek, L. (1969). Plasma insulin and triglyceride levels after hypothalamic lesions in weanling rats. *Am. J. Physiol.*, **216**, 1496–1501.
Frohman, L. A., Muller, E. E. & Cocchi, D. (1973). Central nervous system mediated inhibition of insulin secretion due to 2-deoxyglucose. *Horm. Metab. Res.*, **5**, 21–5.
Fryer, J. H., Moore, N. S., Williams, H. H. & Young, C. M. (1955). A study of the interrelationship of the energy-yielding nutrients, blood glucose levels and subjective appetite in man. *J. Lab. Clin. Med.*, **45**, 684–96.
Galef, B. G., Jr & Henderson, P. W. (1972). Mother's milk: a determinant of the feeding preferences of weanling rat pups. *J. Comp. Physiol. Psychol.*, **78**, 213–19.
Galef, B. G., Jr & Sherry, D. F. (1973). Mother's milk: a medium for transmission of cues reflecting the flavor of mother's diet. *J. Comp. Physiol. Psychol.*, **83**, 374–8.
Ganchrow, J. R., Lieblich, I. & Cohen, E. (1981). Consummatory responses to taste stimuli in rats selected for high and low rates of self stimulation. *Physiol. Behav.*, **27**, 971–6.
Gasnier, A., Gompel, M., Hamon, F. & Mayer, A. (1932). Régulation automatique de l'ingestion et de l'absorption des aliments en fonction de la teneur en eau du régime. *Ann. Physiol.*, **8**, 870–90.
Geary, N., Grotschel, H. & Scharrer, E. (1982). Blood metabolites and feeding during postinsulin hypophagia. *Am. J. Physiol.*, **243**, R304–R311.
Geary, N., Langhans, W. & Scherrer, E. (1981). Metabolic concomitants of glucagon-induced suppression of feeding in the rat. *Am. J. Physiol.*, **241**, R330–R335.
Geiselman, P. H., Rogers, G. H., Jaster, J. P., Martin, J. R. & Novin, D. (1979). System for measurement and data reduction of meal-related parameters. *Physiol. Behav.*, **22**, 397–400.
Geliebter, A. A. (1979). Effects of equicaloric loads of protein, fat and carbohydrate on food intake in the rat and man. *Physiol. Behav.*, **22**, 267–74.

Geliebter, A. A., Torbay, N., Bracco, E. F., Hashim, S. A. & Van Itallie, T. B. (1983). Overfeeding with medium-chain triglyceride diet results in diminished deposition of fat. *Am. J. Clin. Nutr.*, **37**, 1–4.

Gibbs, J., Maddison, S. P. & Rolls, E. T. (1981). Satiety role of the small intestine examined in sham-feeding and rhesus monkeys. *J. Comp. Physiol. Psychol.*, **95**, 1003–15.

Gibson, T., Stimmler, L., Jarrett, R. J., Rutland, P. & Shiu, M. (1975). Diurnal variation in the effects of insulin on blood glucose, nonesterified fatty acids and growth hormone. *Diabetologia*, **11**, 83–8.

Girardier, L., Seydoux, J. & Campfield, L. A. (1976). Control of A and B cells *in vivo* by sympathetic nervous input and selective hyper- or hypo-glycemia in dog pancreas. *J. Physiol. (Lond.)*, **256**, 801–14.

Glenn, J. F. & Erickson, R. P. (1976). Gastric modulation of gustatory afferent activity. *Physiol. Behav.*, **16**, 561–8.

Gli-Ad, I., Udeschini, G., Cocchi, D. & Muller, E. E. (1975). Hyporesponsiveness to glucoprivation during post natal period in the rat. *Am. J. Physiol.*, **229**, 512–17.

Glick, Z. & Modan, M. (1977). Behavioral compensatory responses to continuous duodenal and upper ileal glucose infusion in rats. *Physiol. Behav.*, **19**, 703–5.

Gold, R. M., Jones, A. P., Sawchenko, P. E. & Kapatos, G. (1977). Paraventricular area: critical focus of a longitudinal neurocircuitry mediating food intake. *Physiol. Behav.*, **18**, 1111–20.

Goldman, J. K. & Bernardis, L. L. (1974). Metabolism of glucose, fructose and pyruvate in tissues of weanling rats with hypothalamic obesity. *Horm. Metab. Res.*, **6**, 370–5.

Goldman, J. K., Schnatz, J. D., Bernardis, L. L. & Frohman, L. L. (1972*a*). Effects of ventromedial hypothalamic destruction in rats with preexisting streptozotocin-induced diabetes. *Metabolism*, **21**, 132–6.

(1972*b*). *In vivo* and *in vitro* metabolism in hypothalamic obesity. *Diabetologia*, **8**, 160–4.

Goldstein, R., Hill, S. Y. & Templer, D. L. (1970). Effect of food deprivation on hypothalamic self-stimulation in stimulus-bound eaters and noneaters. *Physiol. Behav.*, **5**, 915–18.

Goldstein, R. & Ripley, C. (1976). The effect of food deprivation on the approach and escape components of VMH self-stimulation. *Physiol. Behav.*, **16**, 131–4.

Gonzalez, M. F. & Novin, D. (1974). Feeding induced by intracranial and intravenously administered 2-DG. *Physiol. Psychol.*, **2**, 326–30.

Green, E., Miller, D. S. & Wynn, V. (1975). Oxygen consumption of obese and anorectic patients. *Proc. Nutr. Soc.*, **34**, 14A–15A.

Green, K. F. & Garcia, J. (1971). Recuperation from illness: flavor enhancement for rats. *Science*, **173**, 749–51.

Grill, H. J. & Norgren, R. E. (1978*a*). Chronically decerebrate rats demonstrate satiation but not bait-shyness. *Science*, **201**, 267.

(1978*b*). The taste reactivity test. I. Mimetic responses to gustatory stimuli in neurologically normal rats. *Brain Res.*, **142**, 263–80.

Grossman, S. P., Daley, D., Halaris, A. E., Collier, T. & Routtenberg, A. (1978). Aphagia and adipsia after preferential destruction of nerve cell bodies in hypothalamus. *Science*, **202**, 537–9.

Grossman, S. P. & Grossman, L. (1982). Iontophoretic injections of kainic acid into rat lateral hypothalamus: effects of ingestive behavior. *Physiol. Behav.*, **29**, 553–60.

Grossman, S. P. & Rechtschaffen, A. (1967). Variations in brain temperature in relation to food intake. *Physiol. Behav.*, **2**, 379–83.

Gurr, M. I., Rothwell, N. J. & Stock, M. J. (1979). Effects of tube-feeding iso-energetic amounts of various dietary lipids on energy balance in the rat. *Proc. Nutr. Soc.*, **38**, A6.

Guttman, N. (1953). Operant conditioning, extinction and periodic reinforcement in relation to concentration of sucrose used as a reinforcing agent. *J. Exp. Psychol.*, **46**, 213–24.

Haessler, H. A. & Crawford, J. D. (1967). Fatty acid composition and metabolic activity of depot fat in experimental obesity. *Am. J. Physiol.*, **213**, 255–61.

Hales, C. N. & Kennedy, G. C. (1964). Plasma glucose, non-esterified fatty acid and insulin concentrations in hypothalamic hyperphagic rats. *Biochem. J.*, **90**, 620–4.

Hall, W. G. (1975). Weaning and growth of artificially reared rats. *Science*, **190**, 1313–15.

Hall, W. G., Cramer, C. P. & Blass, E. M. (1975). Developmental changes in suckling of rat pups. *Nature*, **258**, 318–19.

Hamburg, M. D. (1971). Hypothalamic unit activity and eating behavior. *Am. J. Physiol.*, **220**, 980–5.

Hamilton, C. L. (1964). Rat's preference for high fat diets. *J. Comp. Physiol. Psychol.*, **58**, 459–60.

Han, P. W. (1967). Hypothalamic obesity in rats without hyperphagia. *Trans. N.Y. Acad. Sci.*, **30**, 229–43.

Han, P. W., Feng, L. Y. & Kuo, P. T. (1972). Insulin sensitivity of pair-fed, hyperlipemic, hyperinsulinemic, obese-hypothalamic rats. *Am. J. Physiol.*, **223**, 1206–9.

Hansen, B. C., Jen, K. L. C. & Hagedorn, P. K. (1971). Regulation of food intake in monkeys: response to caloric dilution. *Physiol. Behav.*, **26**, 479–87.

Hansen, B. C., Jen, K. L. C. & Kalnasy, L. W. (1981). Control of food intake and meal patterns in monkeys. *Physiol. Behav.*, **27**, 803–10.

Harris, L. J., Clay, J., Hargreaves, F. J. & Ward, A. (1933). Appetite and choice of diet: the ability of the vitamin B deficient rat to discriminate between diets containing and lacking the vitamin. *Proc. R. Soc. Lond., B.*, **113**, 161–90.

Hasegawa, Y. (1981). Recuperation from lithium-induced illness: flavor enhancement for rats. *Behav. Neural Biol.*, **33**, 252–5.

Havrankova, J., Brownstein, M. & Roth, J. (1981). Insulin and insulin receptors in rodent brain. *Diabetologia*, **20** (Suppl.), 268–73.

Havrankova, J., Roth, J. & Brownstein, M. (1978). Insulin receptors are widely distributed in the central nervous system of the rat. *Nature*, **272**, 827–9.

Herberg, L. J. (1960). Hunger reduction produced by injecting glucose into the lateral ventricle of the rat. *Nature*, **187**, 245–6.

Heron, W. T. & Skinner, B. F. (1937). Changes in hunger during starvation. *Psychol. Rec.*, **1**, 51–60.

Hervey, G. R., Parameswaran, S. V. & Steffens, A. B. (1977). The effects of lateral hypothalamic stimulation in parabiotic rats. *J. Physiol.* (*Lond.*), **266**, 64P.

Hervey, G. R. & Tobin, G. (1982). Brown adipose tissue weight in relation to rat body fat content. *J. Physiol.* (*Lond.*), **330**, 71P–72P.

Hesse, G. W., Hesse, K. A. F. & Catalanotto, F. A. (1979). Behavioral characteristics of rats experiencing chronic deficiency. *Physiol. Behav.*, **22**, 211–16.

Hetherington, A. W. & Ranson, S. W. (1940). Hypothalamic lesions and adiposity in rats. *Anat. Rec.*, **78**, 149–72.

Hirsch, E. (1973). Some determinants of intake and patterns of feeding in the Guinea Pig. *Physiol. Behav.*, **11**, 687–704.

Hirsch, E., Dubose, C. & Jacobs, H. L. (1982). Overeating, dietary selection patterns and sucrose intake in growing rats. *Physiol. Behav.*, **28**, 819–28.

Hoebel, B. G. (1968). Inhibition and disinhibition of self-stimulation and feeding: hypothalamic control and post-ingestional factors. *J. Comp. Physiol. Psychol.*, **66**, 89–100.

Hoebel, B. G. & Teitelbaum, P. (1962). Hypothalamic control of feeding and self-stimulation. *Science*, **135**, 375–7.

(1966). Weight regulation in normal and hypothalamic hyperphagic rats. *J. Comp. Physiol. Psychol.*, **61**, 189–93.

Hoebel, B. G. & Thompson, R. D. (1969). Aversion to lateral hypothalamic stimulation caused by intragastric feeding or obesity. *J. Comp. Physiol. Psychol.*, **68**, 536–43.

Holman, G. L. (1969). Intragastric reinforcement effect. *J. Comp. Physiol. Psychol.*, **69**, 432–41.

Horell, R. I. & Redgrave, P. (1976). The effect of food deprivation on escape from electrical stimulation of the VMH. *Physiol. Psychol.*, **4**, 233–7.

Houpt, K. A. & Epstein, A. N. (1973). Ontogeny of controls of food intake in the rat: GI fill and glucoprivation. *Am. J. Physiol.*, **225**, 58–66.

Houpt, T. R. & Hance, H. E. (1971). Stimulation of food intake in the rabbit and rat by inhibition of glucose metabolism with 2-deoxy-glucose. *J. Comp. Physiol. Psychol.*, **76**, 395–400.

Houpt, K. & Houpt, T. R. (1975). Effect of gastric loads and food deprivation on subsequent food intake in suckling rats. *J. Comp. Physiol. Psychol.*, **88**, 764–72.

Huang, Y. H. & Mogenson, G. J. (1972). Neural pathways mediating drinking and feeding in rats. *Exp. Neurol.*, **37**, 269–86.

Huenemann, R. L. (1972). Food habits of obese and nonobese adolescents. *Postgrad. Med.*, **51**, 99–105.

Hull, C. L., Livingstone, J. R., Rouse, R. A. & Barker, A. N. (1951). True, sham and esophageal feeding as reinforcements. *J. Comp. Physiol. Psychol.*, **44**, 236–45.

Hustvedt, B. E. & Løvø, A. (1972). Correlation between hyperinsulinemia and hyperphagia in rats with ventromedial hypothalamic lesions. *Acta Physiol. Scand.*, **84**, 29–33.

Inoue, S. & Bray, G. A. (1980). Role of the autonomic nervous system in the development of ventromedial hypothalamic obesity. *Brain Res. Bull.*, **5** (Suppl. 4), 119–26.

Inoue, S., Bray, G. A. & Muller, Y. S. (1978). Transplantation of pancreatic β-cells prevents development of hypothalamic obesity in rats. *Am. J. Physiol.*, **235**, E266–E271.

Irsigler, K., Veitl, V., Sigmund, A., Tschegg, E. & Kunz, K. (1979). Calorimetric results in man: energy output in normal and overweight subjects. *Metabolism*, **28**, 1127–32.

Jacobs, H. L. & Sharma, K. N. (1969). Taste versus calories: sensory and metabolic signals in the control of food intake. *Ann. N.Y. Acad. Sci.*, **157** (Art. 2), 1084–125.

Janowitz, H. D. & Grossman, M. I. (1949). Some factors affecting the food intake of normal dogs and dogs with esophagotomy. *Am. J. Physiol.*, **159**, 143–8.

Janowitz, H. D. & Hollander, F. (1955). The time factor in the adjustment of food

intake to varied caloric equipment in the dog: a study of the precision of appetite regulation. *Ann. N.Y. Acad. Sci.*, **63**, 56–67.
Janowitz, H. D. & Ivy, A. C. (1949). Role of blood sugar levels in spontaneous and insulin-induced hunger in man. *J. Appl. Physiol.*, **1**, 643–5.
Jeanningros, R. (1982). Vagal unitary responses to intestinal amino acid infusions in the anesthetized cat: a putative signal for protein induced satiety. *Physiol. Behav.*, **28**, 9–21.
(1983). Effect of intestinal amino acid infusions on hypothalamic single unit activity in the anesthetized cat. *Br. Res. Bull.*, **10**, 15–21.
(1984). Lateral hypothalamic responses to preabsorptive and post-absorptive signals related to amino-acid ingestion. *J. Auton. Nerv. Syst.*, **10**, 261–8.
Jeanningros, R. & Mei, N. (1980). Vagal and splanchnic effects at the level of the ventromedian nucleus of the hypothalamus in the cat. *Brain Res.*, **185**, 239–52.
Jequier, E. (1982/1983). Thermogenesis and human obesity: a review. *Nestlé Res. News*, **27**, 36.
Jiang, C. L. & Hunt, J. N. (1983). The relation between freely chosen meals and body habitus. *Am. J. clin. Nutr.*, **38**, 32–40.
Johnson, D. J., Li, E. T. S., Coscina, D. V. & Anderson, G. H. (1979). Different diurnal rhythms of protein and non-protein energy intake by rats. *Physiol. Behav.*, **22**, 777–80.
Johnson, N. E. & Cark, Z. M. (1947). Environment and food intake in man. *Science*, **105**, 368–79.
Jones, R. G. & Booth, D. A. (1975). Dose–response for 2-deoxy-D-glucose induced feeding and the involvement of peripheral factors. *Physiol. Behav.*, **15**, 85–90.
Jordan, H. A. (1969). Voluntary intragastric feeding: oral and gastric contributions to food intake and hunger in man. *J. Comp. Physiol. Psychol.*, **68**, 498–506.
Jordan, H. A., Wieland, W. F., Zebley, S. P., Stellar, E. & Stunkard, A. J. (1966). Direct measurement of food intake in man: a method for objective study of eating behavior. *Psychosom. Med.*, **28**, 836.
Jouhaneau, J. & Le Magnen, J. (1980). Behavioral regulation of the blood glucose level in rats. *Neurosci. Biobehav. Rev.*, **4** (Suppl. 1), 53–63.
Kakolewski, J. W., Deaux, E., Christensen, J. & Case, B. (1971). Diurnal patterns in water and food intake and body weight changes in rats with hypothalamic lesions. *Am. J. Physiol.*, **221**, 711–18.
Kalogeris, T. J., Reidelberger, R. D. & Mendel, V. E. (1983). Effect of nutrient density and composition of liquid meals on gastric emptying in feeding rats. *Am. J. Physiol.*, **244**, R865–R871.
Kanarek, R. B. & Marks-Kaufman, R. (1979). Developmental aspects of sucrose-induced obesity in rats. *Physiol. Behav.*, **23**, 881–6.
Kasemsri, S., Bernardis, L. L. & Schnatz, J. D. (1972). Fat mobilization in adipose tissue of weanling rats with hypothalamic obesity. *Hormones*, **3**, 97–104.
Kataoka, K., Danbara, H., Sunayashiki, K., Okuno, S., Sorimachi, M., Tanaka, M. & Chikamori, K. (1978). Regional and subcellular distribution of gold in brain of gold thioglucose obese mouse. *Brain Res. Bull.*, **3**, 257–63.
Keesey, R. E., Boyle, P. C., Kemnitz, W. & Mitchel, J. S. (1976). The role of the lateral hypothalamus in determining the body weight set point. In *Hunger: Basic Mechanisms and Clinical Implications*, ed. D. Novin, W. Wyrwicka & G. A. Bray, pp. 243–56. Raven Press: New York.
Kimura, T., Maji, T. & Ashida, K. (1970). Periodicity of food intake and lipogenesis in rats subjected to two different feeding plans. *J. Nutr.*, **100**, 691–7.

Kissileff, H. R. (1968). The effects of water loading and ambient temperature changes on temporal patterns of food and water intake in the rat. In Proceedings of the 3rd International Conference on the Regulation of Food and Water Intake, Haverford, PA.

(1970). Free feeding in normal and 'recovered lateral' rats monitored by a pellet-detecting eatometer. *Physiol. Behav.*, **5**, 163–74.

Kissileff, H. R., Klingsberg, G. & Van Itallie, T. B. (1980). Universal eating monitor for continuous recording of solid or liquid consumption in man. *Am. J. Physiol.*, **238**, R14–R22.

Kissileff, H. R., Thornton, J. & Becker, E. (1982). A quadratic equation adequately describes the cumulative food intake curve in man. *Appetite*, **3**, 255–72.

Knight, N. F., Sanders, C. W., Venables, S. & Zahedi-Asl, S. (1980). Insulin and glucose concentrations before and after feeding in man. *J. Physiol.* (*Lond.*), **308**, 114P.

Kogure, S., Onoda, N. & Takagi, S. F. (1980). Responses of lateral hypothalamic neurons to odours before and during stomach distension in unanesthetized rabbits. In Proceedings of the 28th International Congress of Physiological Science, Budapest.

Kolakowska, L., Larue-Achagiotis, C. & Le Magnen, J. (1984). Effets comparés de la lésion du noyau basolatéral et du noyau latéral de l'amygdale sur la néophobie et l'aversion gustative conditionnée chez le rat. *Physiol. Behav.*, **32**, 647–51.

Koopmans, H. S. (1981). Peptides as satiety agents: the behavioral evaluation of their effects on food intake. In *Gut Hormones*, ed. S. R. Bloom & J. M. Polak, pp. 464–70. Churchill Livingstone: London.

Kral, J. G., Powley, T. L. & Brooks, C. McC. (ed.) (1983). *Vagal Nerve Function: Behavioral and Methodological Considerations.* Elsevier: Amsterdam.

Kraly, F. S. (1980). Decreased satiety potency of cholecystokinin in the night phase of the rat's diurnal cycle. Proceedings of the 7th International Conference on Food and Fluid Intake, Warsaw.

(1981). Pre-gastric stimulation and cholecystokynin are not sufficient for the meal size–intermeal interval correlation in the rat. *Physiol. Behav.*, **27**, 457–62.

Kraly, F. S. & Blass, E. M. (1976). Increased feeding in rats in a low ambient temperature. In *Hunger: Basic Mechanisms and Clinical Implications*, ed. D. Novin, W. Wyrwicka & G. A. Bray, pp. 77–88. Raven Press: New York.

Kraly, F. S., Carty, W. J. & Smith, G. P. (1978). Effect of pregastric food stimuli on meal size and intermeal interval in the rat. *Physiol. Behav.*, **20**, 779–84.

Kraly, F. S. & Gibbs, J. (1980). Vagotomy fails to block the satiating effect of food in the stomach. *Physiol. Behav.*, **24**, 1007–10.

Kraly, F. S. & Smith, G. P. (1978). Combined pregastric and gastric stimulation by food is sufficient for normal meal size. *Physiol. Behav.*, **21**, 405–8.

Kratz, C. M., Levitsky, D. A. & Lustick, S. (1978). Differential effects of quinine and sucrose octo-acetate on food intake in the rat. *Physiol. Behav.*, **20**, 665–8.

Krauss, R. M. & Mayer, J. (1965). Influence of protein and amino-acids on food intake in the rat. *Am. J. Physiol.*, **209**, 479–83.

Kristal, M. B. & Wampler, R. S. (1973). Food and water intake prior to parturition in the rat. *Physiol. Psychol.*, **1**, 297.

Kumon, A., Hara, T. & Tahakashi, A. (1976). Effects of catecholamines on the lipolysis of two kinds of fat cells from adult rabbit. *J. Lipid. Res.*, **17**, 559–64.

Langfeld, H. S. (1914). The psychophysiology of a long prolonged fast. *Psychol. Monogr.*, **16**, 61–2.

Langhans, W., Geary, N. & Scharrer, E. (1982). Liver glycogen content decreases during meals in rats. *Am. J. Physiol.*, **243**, R450–R453.

Larue, C. (1973). Les mécanismes du contrôle olfactif de la prise alimentaire chez le rat blanc. Thèse de Doctorat d'Etat Sciences Naturelles, University of Paris.

Larue, C. & Le Magnen, J. (1972). The olfactory control of meal pattern in rats. *Physiol. Behav.*, **9**, 817–21.

(1973). Effets de l'interruption des voies olfacto-hypothalamiques sur la séquence alimentaire du rat. *J. Physiol. (Paris)*, **66**, 699–713.

Larue-Achagiotis, C. & Le Magnen, J. (1979). The different effects of continuous night and day-time insulin infusion on the meal pattern of normal rats: comparison with the meal pattern of hyperphagic hypothalamic rats. *Physiol. Behav.*, **22**, 435–40.

(1982). Effects of short-term nocturnal and diurnal food deprivation on subsequent feeding in intact and VMH lesioned rats: relations to blood glucose level. *Physiol. Behav.*, **16**, 245–8.

(1985*a*). Effect of long-term intravenous insulin infusion on body weight on food intake in intravenous versus intraperitoneal routes. *Appetite*, in press.

(1985*b*). Feeding rate and responses to food deprivation as a function of fast-induced hypoglycemia. *Behav. Neurosci.*, in press.

Lawrence, D. H. & Mason, W. A. (1955). Intake and weight adjustment in rats to changes in feeding schedules. *J. Comp. Physiol. Psychol.*, **48**, 43–6.

Leathwood, P. & Arimanana, L. (1980). Patterns of food intake in rats offered a choice of high and low protein diets. In Proceedings of the 7th International Conference of the Physiology of Food and Fluid Intake.

Leibowitz, S. F., Hammer, N. J. & Chang, K. (1981). Hypothalamic para-ventricular nucleus lesions produce overeating and obesity in the rat. *Physiol. Behav.*, **27**, 1031–40.

Le Magnen, J. (1953*a*). Activité de l'insuline sur la consommation spontanée des solutions sapides. *C.r. Soc. Biol. (Paris)*, **147**, 1753–7.

(1953*b*). Nouvelles données sur le processus de régulation des consommations hydrique et saline chez le rat blanc, *C. r. Soc. Biol. (Paris)*, **147**, 1675–7.

(1954). Le processus de discrimination par le rat blanc des stimuli sucrés alimentaires et non alimentaires. *J. Physiol. (Paris)*, **46**, 414–18.

(1956). Hyperphagie provoquée chez le rat blanc par l'altération du mécanisme de satiété périphérique. *C. r. Soc. Biol.*, **150**, 32–4.

(1959*a*). Effet des administrations post-prandiales de glucose sur l'établissement des appétits. *C. r. Soc. Biol.*, **153**, 212–15.

(1959*b*). Etude d'un phénomène d'appétit provisionnel. *C. r. Acad. Sci. (Paris)*, **249**, 2400–2.

(1960). Effets d'une pluralité de stimuli alimentaires sur le déterminisme quantitatif de l'ingestion chez le rat blanc. *Arch. Sci. Physiol.*, **14**, 411–19.

(1969). Peripheral and systemic actions of food in the caloric regulation of intake. *Ann. NY Acad. Sci.*, **268**, 3107–10.

(1976). Interactions of glucostatic and lipostatic mechanisms in the regulatory control of feeding. In *Hunger: Basic Mechanisms and Clinical Implications*, ed. D. Novin, W. Wyrwicka & G. A. Bray, pp. 89–101. Raven Press: New York.

(1977*a*). Sweet preference and the sensory control of intake. In *Taste and Development: The Genesis of Sweet Preference*, ed. J. M. Weinffenbach, pp. 355–62. US Dept. of Health, Education & Welfare: Bethesda, Maryland.

(1977*b*). Hunger and food palatability in the control of feeding behaviour. In *Food Intake and Chemical Senses*, ed. Y. Katsuki, M. Sato, S. Takagi & Y. Oomura, pp. 263–80. University of Tokyo Press: Tokyo.

(1981). The metabolic basis of dual periodicity of feeding in rats. *Behav. Brain Sci.*, **4**, 561–607.

(1983). Body energy balance and food intake: a neuroendocrine regulatory mechanism. *Physiol. Rev.*, **63**, 314–87.

(1984). Is regulation of body weight elucidated? *Neurosci. Biobehav. Rev.*, **8**, 515–22.

Le Magnen, J. & Devos, M. (1970). Metabolic correlates of the meal onset in the free food intake of rats. *Physiol. Behav.*, **5**, 805–14.

(1980*a*). Parameters of the meal pattern in rats: their assessment and physiological significance. *Neurosci. Biobehav. Rev.*, **4** (Suppl. 1), 1–11.

(1980*b*). Variations of meal-to-meal liver glycogen in rats. *Neurosci. Biobehav. Rev.*, **4** (Suppl. 1), 29–32.

(1982). Daily body energy balance in rats. *Physiol. Behav.*, **29**, 807–11.

(1984). Meal to meal energy balance. *Physiol. Behav.*, **32**, 39–44.

Le Magnen, J., Devos, M., Gaudilliere, J. P., Louis-Sylvestre, J. & Tallon, S. (1973). Role of a lipostatic mechanism in regulation by feeding of energy balance in rats. *J. Comp. Physiol. Psychol.*, **84**, 1–23.

Le Magnen, J., Devos, M. & Larue-Achagiotis, C. (1980*a*). Food deprivation induced parallel changes in blood glucose plasma free fatty acids and feeding during two parts of the diurnal cycle in rats. *Neurosci. Biobehav. Rev.*, **4** (Suppl. 1), 17–23.

Le Magnen, J. & Marfaing-Jallat, P. (1984). Further study of induced behavioral dependence on ethanol in rats. *Alcohol*, **1**, 269–73.

Le Magnen, J., Marfaing-Jallat, P., Miceli, D. & Devos, M. (1980*b*). Pain-modulating and reward systems: a single brain mechanism? *Pharmacol. Biochem. Behav.*, **12**, 707–9.

Le Magnen, J. & Tallon, S. (1963). Enregistrement et analyse préliminaire de la 'périodicité alimentaire spontanée' chez le rat blanc. *J. Physiol.* (*Paris*), **55**, 286–97.

(1966). La périodicité spontanée de la prise d'aliments *ad libitum* du rat blanc. *J. Physiol.* (*Paris*), **58**, 323–49.

(1968*a*). L'effet du jeûne préalable sur les caractéristiques temporelles de la prise d'aliments chez le rat. *J. Physiol.* (*Paris*), **60**, 143–54.

(1968*b*). Préférence alimentaire du jeune rat induite par l'allaitement maternel. *C. r. Soc. Biol.*, **162**, 387–90.

Le Marchand, Y., Freychet, P. & Jeanrenaud, B. (1978). Longitudinal study on establishment of insulin resistance in hypothalamic obese mice. *Endocrinology*, **102**, 74–85.

Leung, P. M. & Rogers, Q. R. (1969). Food intake: regulation by plasma amino acid pattern. *Life Sci.*, **8**, 1–9.

(1970). Effect of amino acid imbalance and deficiency on food intake of rats with hypothalamic lesions. *Nutr. Rep. Int.*, **1**, 1–10.

(1971). Importance of prepyriform cortex in food intake response of rats to amino acids. *Am. J. Physiol.*, **198**, 1315–18.

(1973). Effect of amygdaloid lesions on dietary intake of disproportionate amounts of amino acids. *Physiol. Behav.*, **11**, 221–6.

Levin, R. & Stern, J. M. (1975). Maternal influences on ontogeny of suckling and feeding rhythms in the rat. *J. Comp. Physiol. Psychol.*, **89**, 711–21.

Levitsky, D. A., Faust, I. & Glassman, M. (1976). The ingestion of food and the recovery of body weight following fasting in the naive rats. *Physiol. Behav.*, **17**, 575–80.

Lewis, M. (1964). Behavior resulting from calcium deprivation in parathyroidectomized rats. *J. Comp. Physiol. Psychol.*, **57**, 348–52.

Li, E. T. & Anderson, G. H. (1982). Meal composition influences subsequent food selection in the young rat. *Physiol. Behav.*, **29**, 779–84.

Lieblich, I., Cohen, E., Ganchrow, J. R., Blass, E. M. & Bergman, F. (1983). Morphine tolerance in genetically selected rats induced by chronically elevated saccharine intake. *Science*, **221**, 871–3.

Liebling, D. S., Eisner, J. D., Gibbs, J. & Smith, G. P. (1975). Intestinal satiety in rats. *J. Comp. Physiol. Psychol.*, **89**, 955–66.

Likuski, H. J., Debons, A. F. & Cloutier, R. J. (1967). Inhibition of gold thioglucose-induced hypothalamic obesity by glucose analogues. *Am. J. Physiol.*, **212**, 669–76.

Lipsitt, L. P. (1977). Taste in human neonates: its effects on sucking and heart rate. In *Taste and Development: The Genesis of Sweet Preference*, ed. J. M. Weiffenbach, pp. 125–45. US Dept of Health, Education & Welfare: Bethesda, Maryland.

Louis-Sylvestre, J. (1976). Preabsorptive insulin release and hypoglycemia in rats. *Am. J. Physiol.*, **230**, 56–60.

(1978*a*). Metabolic and feeding patterns in rats with truncular vagotomy or with transplanted beta cells. *Am. J. Physiol.*, **235**, E119–E125.

(1978*b*). Relationship between two stages of prandial insulin release in rats. *Am. J. Physiol.*, **235**, E103–E111.

(1983*a*). Validation of tests of completeness of vagotomy in rats. In *Vagal Nerve Function: Behavioral and Methodological Considerations*, ed. J. G. Kral, T. L. Powley & C. McC. Brooks, pp. 301–14. Elsevier: Amsterdam.

(1983*b*). Phase céphalique de secrétion d'insuline et variété des aliments au cours du repas chez le rat. *Reprod. Nutr. Develop.*, **23**, 351–6.

Louis-Sylvestre, J. & Le Magnen, J. (1980*a*). A fall in blood glucose level precedes meal onset in free-feeding rats. *Neurosci. Biobehav. Res.*, **4** (Suppl. 1), 13–16.

(1980*b*). Palatability and preabsorptive insulin release. *Neurosci. Biobehav. Rev.*, **4** (Suppl. 1), 43–6.

(1984). Dietary versus sensory factors in cafeteria-induced obesity. *Physiol. Behav.*, **32**, 901–6.

Louis-Sylvestre, J., Servant, J. M. & Molimard, R. (1979). Effet de l'anastomose portocave sur la séquence alimentaire du rat. *J. Physiol. (Paris)*, **75**, A92.

Louis-Sylvestre, J., Servant, J. M., Molimard, R. & Le Magnen, J. (1980). Effect of liver denervation on the feeding pattern of rats. *Am. J. Physiol.*, **239**, R66–R70.

Løvø, A. & Hustvedt, B. E. (1973). Correlation between altered acetate utilization and hyperphagia in rats with ventromedial hypothalamic lesions. *Metabolism*, **22**, 1459–66.

Lowell, B. B., Wade, G., Gray, J. M., Gold, R. M. & Petrulavage, J. (1980). Adipose tissue lipoprotein lipase activity in rats with obesity inducing hypothalamic knife cuts. *Physiol. Behav.*, **25**, 113–16.

Lundbaeck, K. & Stevenson, J. A. F. (1947). Reduced carbohydrate intake after fat feeding in normal rats and rats with hypothalamic hyperphagia. *Am. J. Physiol.*, **151**, 530–7.

Luria, Z. (1953). Behavioral adjustment to thiamine deficiency in albino rats. *J. Comp. Physiol. Psychol.*, **46**, 358–62.

Lytle, L. D., Moorcroft, W. H. & Campbell, B. A. (1971). Ontogeny of amphetamine anorexia and insulin hyperphagia in the rat. *J. Comp. Physiol. Psychol.*, **77**, 388–93.

McCracken, K. J. & Barr, H. G. (1982). Energy balance and body fat changes in young 'cafeteria'-fed rats kept at 24 °C. *J. Physiol.* (*Lond.*), **330**, 69P–70P.

McDonald, G. B., Saunders, D. R., Weidman, M. & Fisher, L. (1980). Portal venous transport of long-chain fatty acids absorbed from rat intestine. *Am. J. Physiol.*, **239**, G141–G150.

McGinty, D., Epstein, A. N. & Teitelbaum, P. (1965). The contribution of oropharyngeal sensations to hypothalamic hyperphagia. *Anim. Behav.*, **13**, 413–18.

McHugh, P. R. & Moran, T. H. (1978). The accuracy of regulation of caloric ingestion in the rhesus monkey. *Am. J. Physiol.*, **235**, R29–R34.

McHugh, P. R., Moran, T. H. & Barton, G. N. (1975). Satiety: a graded behavioral phenomenon regulating caloric intake. *Science*, **190**, 167–9.

MacKay, E. M., Callaway, J. W. & Barnes, R. (1940). Hyperalimentation in normal animals produced by protamine insulin. *J. Nutr.*, **20**, 59–66.

MacLeod, P. (1971). Structure and function of higher olfactory centers. In *Handbook of Sensory Physiology*, vol. IV *Chemical Senses*, Part 1 *Olfaction*, ed. L. M. Beidler, pp. 182–204. Springer Verlag: Berlin.

MacNeil, D. (1974). Lateral hypothalamic self-stimulation; excess of body weight. *Physiol. Psychol.*, **2**, 51–2.

Maggio, C. A. & Koopmans, H. S. (1982). Food intake after intragastric meals of short-, medium- or long-chain triglyceride. *Physiol. Behav.*, **28**, 921–6.

Maji, T. & Ashida, K. (1978). Effect of alternate high carbohydrate and high fat diets on glucose tolerance. *Nutr. Rep. Int.*, **17**, 637–44.

Malherbe, C., De Gasparo, M., De Hertogh, R. & Hoet, J. J. (1969). Circadian variations on blood sugar and plasma insulin levels in man. *Diabetologia*, **5**, 397–404.

Mandenoff, A., Lenoir, T. & Apfelbaum, M. (1982). Tardy occurrence of adipocyte hyperplasia in cafeteria-fed rat. *Am. J. Physiol.*, **242**, R349–R351.

Marcilloux, J. C. (1980). Comportement alimentaire de l'oie de race landaise: importance du noyau ventro-médian de l'hypothalamus dans sa régulation. Thèse de Doctorat d'Université, Université Pierre et Marie Curie, Paris VI.

Margules, D. L. & Olds, J. (1962). Identical 'feeding' and 'rewarding' systems in the lateral hypothalamus of rats. *Science*, **135**, 374–5.

Marshall, J. F. & Teitelbaum, P. (1974). Further analysis of sensory inattention following lateral hypothalamic damage in rats. *J. Comp. Physiol. Psychol.*, **86**, 375–95.

Marshall, J. F., Turner, B. H. & Teitelbaum, P. (1971). Sensory neglect produced by lateral hypothalamic damage. *Science*, **174**, 523–5.

May, K. K. & Beaton, J. R. (1968). Hyperphagia in the insulin-treated rat. *Proc. Soc. Exp. Biol. Med.*, **127**, 1201–4.

Mayer, J. (1953). Glucostatic mechanisms of regulation of food intake. *New England J. Med.*, **249**, 13–16.

Mayer, J. & Marshall, N. B. (1956). Specificity of gold-thioglucose for ventromedial hypothalamic lesions and obesity. *Nature*, **178**, 1399–1400.

Mayer, J., Marshall, N. B., Vitale, J. J., Christensen, J. H., Mashayekhi, M. B. & Stare, F. J. (1954). Exercise, food intake and body weight in normal rats and genetically obese adult mice. *Am. J. Physiol.*, **177**, 544–8.

Mayer, J., Monello, L. F. & Seltzer, C. (1965). Hunger and satiety sensations in man. *Postgrad. Med.*, **37**, A97–A102.

Mei, N. (1978). Vagal glucoreceptors in the small intestine of the cat. *J. Physiol.* (*Lond.*), **282**, 485–506.

Mei, N., Arlhac, A. & Boyer, A. (1981). Nervous regulation of insulin by the intestinal vagal glucoreceptors. *J. Auton. Nerv. Syst.*, **4**, 351–63.

Melinkoff, S. M., Frankland, M. & Greisel, M. (1959). Effects of amino-acids and glucose ingestion on arterial venous blood sugar on appetite. *J. Appl. Physiol.*, **9**, 85–7.

Meliza, L. L., Leung, P. M. & Rogers, Q. R. (1981). Effect of anterior prepyriform and medial amygdaloid lesions on acquisition of taste-avoidance and response to dietary amino-acid imbalance. *Physiol. Behav.*, **26**, 1031–7.

Melnyk, R. B., Mrosovsky, N. & Martin, J. M. (1983). Spontaneous obesity and weight loss: insulin binding and lipogenesis in the dormouse. *Am. J. Physiol.*, **245**, R396–R402.

Mendelson, J. (1969). Lateral hypothalamic stimulation: inhibition of aversive effects by feeding, drinking and gnawing. *Science*, **166**, 1431–3.

Miller, N. E. (1955). Short comings of food consumption as a measure of hunger: results from other behavioral techniques. *Ann. N.Y. Acad. Sci.*, **63**, 141–3.

Miselis, R. R. & Epstein, A. N. (1975). Feeding induced by intracerebroventricular 2-deoxy-D-glucose in the rat. *Am. J. Physiol.*, **229**, 1438–47.

Mook, D. G. (1963). Oral and postingestional determinants of the intake of various solutions in rats with esophageal fistulas. *J. Comp. Physiol. Psychol.*, **56**, 645–59.

Mook, D. G., Culberson, R., Gelbart, R. J. & McDonald, K. (1983). Oropharyngeal control of ingestion in rats: acquisition of sham-drinking patterns. *Behav. Neurosci.*, **97**, 574–84.

Mora, F., Rolls, E. T. & Burton, M. J. (1976). Modulation during learning of the responses of neurons in the lateral hypothalamus to the sight of food. *Exp. Neurol.*, **53**, 508–19.

Morath, M. (1974). The 4 hr feeding rhythm of the baby as a free running endogenously regulated rhythm. *Int. J. Chronobiol.*, **2**, 39–46.

Morley, K. E., Levine, A. S., Yim, G. K. & Lowy, M. T. (1983). Opioid modulation of appetite. *Neurosci. Biobehav. Rev.*, **7**, 281–305.

Mrosovsky, N. & Sherry, D. F. (1980). Animal anorexias. *Science*, **207**, 837–42.

Mugford, R. A. (1977). External influences of the feeding in carnivores. In *Chemical Senses & Nutrition*, ed. M. R. Kare & O. Maller, pp. 25–54. Academic Press: New York

Myers, R. D., Casaday, G. & Holman, R. B. (1967). A simplified intracranial cannula for chemical stimulation or long-term infusion of the brain. *Physiol. Behav.*, **2**, 87–8.

Nachman, M. (1963). Learned aversion to the taste of lithium chloride and generalization to other salts. *J. Comp. Physiol. Psychol.*, **56**, 343–9.

Nagai, K., Yamamoto, H. & Nakagawa, H. (1982). Time-dependent hyperglycemic actions of centrally administered 2-deoxy-D-glucose, D-mannitol & D-glucose. *Biomed. Res.*, **3**, 288–93.

Naito, C., Yoshitoshi, Y., Higo, K. & Ookawa, H. (1973). Effects of long-term administration of 2-deoxy-D-glucose on food intake and weight gain in rats. *J. Nutr.*, **103**, 730–7.

Natelson, B. H. & Bonbright, J. C. (1978). Patterns of eating and drinking in the monkeys when food and water are free and when they are earned. *Physiol. Behav.*, **21**, 201–14.

Newman, J. C. & Booth, D. A. (1981). Gastrointestinal and metabolic consequences of a rat's meal on maintenance diet *ad libitum*. *Physiol. Behav.*, **27**, 929–40.

Nicolaïdis, S. & Rowland, N. (1976). Metering of intravenous versus oral nutrients and regulation of energy balance. *Am. J. Physiol.*, **231**, 661–8.

Nicolaïdis, S., Rowland, N., Meile, M. J., Marfaing-Jallat, P. & Pesez, A. (1974). A flexible technique for long-term infusions in unrestrained rats. *Pharmacol. Biochem. Behav.*, **2**, 131–6.

Niijima, A. (1975). The effect of 2-deoxy-D-glucose and D-glucose on the efferent discharge rate of sympathetic nerves. *J. Physiol.* (*Lond.*), **251**, 231–43.

Nikoletesas, M. M. (1980). Food intake in the exercising rat: a brief review. *Neurosci. Biobehav. Rev.*, **4**, 265–7.

Nishizawa, Y. & Bray, G. A. (1978). Ventromedial hypothalamic lesions and mobilization of fatty acids. *J. Clin. Invest.*, **61**, 714–21.

Norgren, R. (1976). Taste pathways to hypothalamus and amygdala. *J. Comp. Neurol.*, **166**, 17–30.

Novin, D., Rogers, R. C. & Hermann, G. (1981). Visceral afferent and efferent connections in the brain. *Diabetologia*, **20** (Suppl.), 331–6.

Novin, D., Sanderson, J. & Gonzalez, M. (1979). Feeding after nutrient infusions: effects of hypothalamic lesions and vagotomy. *Physiol. Behav.*, **22**, 107–14.

Nowlis, G. H. (1973). Taste-elicited tongue movements in human newborn infants: an approach to palatability. In *Oral Sensation and Perception*, ed. J. F. Bosma, pp. 292–303. US Dept of Health Education and Welfare: Bethesda, Maryland.

Ono, T., Nishino, H., Sasaki, K., Fukuda, M. & Muramoto, K. I. (1980). Role of the lateral hypothalamus and the amygdala in feeding behaviour. *Brain Res. Bull.*, **5** (Suppl. 4), 143–50.

(1981). Monkey lateral hypothalamic neuron response to sight of food and during bar press and ingestion. *Neurosci. Lett.*, **21**, 99–104.

Oomura, Y. (1976). Significance of glucose, insulin and free fatty acid on the hypothalamic feeding and satiety neurons. In *Hunger*: *Basic Mechanisms and Clinical Implications*, ed. D. Novin, W. Wyrwicka & G. Bray, pp. 145–58. Raven Press: New York.

Oomura, Y. & Kita, H. (1981). Insulin acting as a modulator of feeding through the hypothalamus. *Diabetologia*, **20** (Suppl.), 290–8.

Opsahl, C. A. (1977). Sympathetic nervous system involvement in the lateral hypothalamic lesion syndrome. *Am. J. Physiol.*, **232**, R128–136.

Pager, J., Giachetti, I., Holley, A. & Le Magnen, J. (1972). A selective control of olfactory bulb electrical activity in relation to food deprivation and satiety in rats. *Physiol. Behav.*, **9**, 573–80.

Pager, J. & Royet, J. P. (1976). Some effects of conditioned aversion on food intake and olfactory bulb electrical responses in the rat. *J. Comp. Physiol. Psychol.*, **90**, 67–77.

Panerai, A. E., Olgiati, V. R., Udeschini, G., Cocchi, D., Pecile, A. & Muller, E. E. (1975). Hyperglycemia and inhibition of insulin secretion by 2-deoxy-D-glucose in rats with hypothalamic lesions. *Pharmacol. Res. Commun.*, **7**, 133–42.

Panksepp, J. (1971). Is satiety mediated by the ventromedial hypothalamus? *Physiol. Behav.*, **7**, 381–4.

(1972). Hypothalamic radioactivity after intragastric glucose-^{14}C in rats. *Am. J. Physiol.*, **223**, 396–401.

(1973). Reanalysis of feeding patterns in the rat. *J. Comp. Physiol. Psychol.*, **82**, 78–94.

Panksepp, J., Pollack, A., Krost, K. P., Mieker, R. & Ritter, M. (1975). Feeding in response to repeated protamine zinc insulin injections. *Physiol. Behav.*, **14**, 487–93.

Parameswaran, S. V., Steffens, A. B., Hervey, G. R. & De Ruiter, L. (1977). Involvement of a humoral factor in regulation of body weight in parabiotic rats. *Am. J. Physiol.*, **132**, R150–R157.

Penicaud, L., Larue-Achagiotis, C. & Le Magnen, J. (1983). Endocrine basis for weight gain after fasting or VMH lesion. *Am. J. Physiol.*, **245**, E246–E252.

Penicaud, L. & Le Magnen, J. (1980*a*). Aspects of the neuroendocrine bases of the diurnal metabolic cycle in rats. *Neurosci. Biobehav. Rev.*, **4** (Suppl. 1), 39–42.

(1980*b*). Recovery of body weight following starvation or food restriction in rats. *Neurosci. Biobehav. Rev.*, **4** (Suppl. 1), 47–52.

Perez-Zahler, L. & Harper, A. E. (1972). Effects of dietary amino-acid pattern on food preference behavior of rats. *J. Comp. Physiol. Psychol.*, **81**, 155–62.

Perkins, M. N., Rothwell, N. J., Stock, M. J. & Stone, T. W. (1981). Activation of brown adipose tissue thermogenesis by electrical stimulation of the ventromedial hypothalamus (abstr.). *J. Physiol.* (*Lond.*), **310**, 32P–33P.

Petersen, S. (1978). Feeding, blood glucose and plasma insulin of mice at dusk. *Nature*, **275**, 647–9.

Pierson, A. & Le Magnen, J. (1970). Study of food textures by recording of chewing and swallowing movements. *J. Text. Stud.*, **1**, 327–37.

Pliner, P. (1982). The effects of mere exposure on liking for edible substances. *Appetite*, **3**, 282–90.

Pliner, P., Rozin, P., Cooper, M. & Woody, G. (1985). Role of medicinal context and specific postingestional effects in the acquisition of liking for tastes. *Appetite*, in press.

Pokrovsky, V. & Le Magnen, J. (1963). Réalisation d'un dispositif d'enregistrement graphique continu et automatique de la consommation alimentaire du rat blanc. *J. Physiol.* (*Paris*), **55**, 318–19.

Porikos, K. P., Hesser, M. F. & Van Itallie, T. B. (1982). Caloric regulation in normal-weight men maintained on a palatable diet of conventional foods. *Physiol. Behav.*, **29**, 293–300.

Portet, R. (1981). Lipid biochemistry in the cold acclimated rat. *Comp. Biochem. Physiol. B*, **70**, 679–88.

Poschel, B. P. (1968). Do biological reinforcers act via the self stimulation areas of the brain? *Physiol. Behav.*, **3**, 53–60.

Powley, T. L. & Laughton, W. (1981). Neural pathways, involved in the hypothalamic integration of autonomic responses. *Diabetologia*, **20** (Suppl.), 378–87.

Powley, T. L. & Opsahl, C. A. (1974). Ventromedial hypothalamic obesity abolished by subdiaphragmatic vagotomy. *Am. J. Physiol.*, **226**, 25–33.

Powley, T. L. & Plocher, T. A. (1980). Hypophysectomy blocks the weight gain and obesity produced by gold-thioglucose lesions. *Behav. Neural. Biol.*, **28**, 300–18.

Randle, P. J. (1965). The glucose fatty acid cycle. In *On the Nature and Treatment of Diabetes*, ed. B. S. Leibel & G. A. A. Wrenshall, pp. 361–7. Excerpta Medica: Amsterdam.

Reidelberger, R. D., Kalogeris, T. J., Leug, P. M. B. & Mendel, V. E. (1983). Postgastric satiety in the sham-feeding rat. *Am. J. Physiol.*, **244**, R872–R881.

Ricardo, J. A. & Koh, E. T. (1978). Anatomical evidence of direct projections from the nucleus of the solitary tract to the hypothalamus, amygdala and other forebrain structures in the rat. *Brain Res.*, **153**, 1–26.
Richter, C. P. (1936). Increased salt appetite in adrenalectomized rats. *Am. J. Physiol.*, **115**, 155–61.
Richter, C. P. & Helfick, S. (1943). Decreased phosphorous appetite of parathyroidectomized rats. *Endocrinology*, **33**, 339–52.
Richter, C. P., Holt, L. E. & Barelare, B. (1937). Vitamin B craving in rats. *Science*, **86**, 354.
Richter, C. P., Holt, L. E., Barelare, B. & Hawkes, C. D. (1938). Changes in fat, carbohydrate and protein appetite in vitamin B deficiency. *Am. J. Physiol.*, **124**, 596–702.
Richter, C. P. & Schmidt, E. C. (1941). Increased fat and decreased carbohydrate appetite of pancreasectomized rats. *Endocrinology*, **28**, 179–92.
Rietveld, W. J., Ten Hoor, F., Kooij, M. & Flory, W. (1979). Maintenance of 24 hour eating rhythmicity during gold thioglucose induced hypothalamic hyperphagia in rats. *Physiol. Behav.*, **22**, 549–54.
Riley, J. N., Card, J. P. & Moore, R. Y. (1981). A retinal projection of the lateral hypothalamus in the rat. *Cell Tiss. Res.*, **214**, 257–69.
Ritter, R. C. & Slusser, P. (1980). 5-Thio-D-glucose causes increased feeding and hyperglycemia in the rat. *Am. J. Physiol.*, **238**, E141–E144.
Rodgers, W. L., Epstein, A. N. & Teitelbaum, P. (1965). Lateral hypothalamic aphagia: motor failure or motivational deficit? *Am. J. Physiol.*, **208**, 334–42.
Rogers, P. J. & Blundell, J. E. (1980). Investigation of food selection and meal parameters during the developments of dietary induced obesity. *Appetite*, **1**, 85.
Rogers, R. C., Kita, H., Butcher, L. L. & Novin, D. (1980). Afferent projections to the dorsal motor nucleus of the vagus. *Brain Res. Bull.*, **5**, 365–74.
Rolland-Cachera, M. F., Deheger, M., Guilloud-Bataille, M., Pequignot, F., Pomeau, Y. & Roche, R. (1983). Relations entre alimentation et corpulence chez l'enfant. *Cahiers Nutr. Dietet.*, **18**, 310–11.
Rolls, B. J. & Rowe, E. A. (1977). Dietary obesity: permanent changes in body weight. *J. Physiol.* (*Lond.*), **272**, 2P.
Rolls, B. J., Van Duijvenvoorde, P. M. & Rowe, E. A. (1983). Variety in the diet enhances intake in a meal and contributes to the development of obesity in the rat. *Physiol. Behav.*, **31**, 21–8.
Rolls, E. T., Burton, M. J. & Mora, F. (1976). Hypothalamic neuronal responses associated with the sight of food. *Brain Res.*, **111**, 53–66.
Rolls, E. T. & Rolls, B. J. (1973). Food preferences after lesion in the basolateral lesion of amygdala. *J. Comp. Physiol. Psychol.*, **83**, 248–59.
(1981). Brain mechanisms involved in feeding. In *The Psychobiology of Human Food Selection*, ed. L. M. Barker, pp. 33–62. AVI Publ. Co., Westport, Conn.
Rothwell, N. J., Saville, M. E. & Stock, M. J. (1982). Sympathetic and thyroid influences on metabolic rate in fed, fasted and refed rats. *Am. J. Physiol.*, **243**, R339–R346.
Rothwell, N. J. & Stock, M. J. (1979). A role for brown adipose tissue in diet-induced thermogenesis. *Nature*, **281**, 31–4.
(1982). Energy expenditure derived from measurements of oxygen consumption and energy balance in hyperphagic cafeteria-fed rats. *J. Physiol.* (*Lond.*), **324**, 59P.
Rowland, N. (1977). Fragmented behaviour sequences in rats with lateral

hypothalamic lesions: an alternative reason for intrameal prandial drinking. *J. Comp. Physiol. Psychol.*, **91**, 1039–55.
Rowland, N., Meile, M. J. & Nicolaïdis, S. (1973). Action inadéquate des apports parentéraux sur l'insulinosecrétion et sur le contrôle du comportement alimentaire chez le rat. *C. r. Acad. Sci. (Paris)*, **277**, 1783–6.
(1975). Metering of intravenously infused nutrients in VMH lesioned rats. *Physiol. Behav.*, **15**, 443–8.
Royle, G. T., Wolfe, R. R. & Burke, J. F. (1982). Glucose and fatty acid kinetics in fasted rats: effects of previous protein intake. *Metabolism*, **31**, 279–83.
Rozin, P. (1965). Specific hunger for thiamine: recovery from deficiency and thiamine preference. *J. Comp. Physiol. Psychol.*, **59**, 98–101.
(1968). Are carbohydrate and protein intakes separately regulated? *J. Comp. Physiol. Psychol.*, **65**, 23–9.
Rozin, P., Gruss, L. & Berk, G. (1979). Reversal of innate aversions: attempts to induce a preference for Chili pepper in rats. *J. Comp. Physiol. Psychol.*, **93**, 1001–14.
Russek, M. (1971). Hepatic receptors and the neurophysiological mechanisms controlling feeding behavior. *Neurosci. Res.*, **4**, 213–82.
Sanahuja, J. C. & Harper, A. E. (1962). Effect of amino acid imbalance on food intake and preference. *Am. J. Physiol.*, **202**, 165–8.
Sanderson, J. D. & Vanderweele, D. A. (1975). Analysis of feeding patterns in normal and vagotomized rabbits. *Physiol. Behav.*, **15**, 357–64.
Sandrew, B. B. & Mayer, J. (1973). Hyperphagia induced by intrahypothalamic implants of mercury thioglucose. *Physiol. Behav.*, **10**, 1061–6.
Saper, C. B., Loewy, A. D., Swanson, L. W. & Cowan, W. M. (1976). Direct hypothalamo-autonomic connections. *Brain Res.*, **117**, 305–12.
Sawchenko, P. E., Gold, R. M. & Alexander, J. (1981). Effects of selective vagotomies on knife-cut induced hypothalamic obesity: differential results on a lab chow vs high-fat diets. *Physiol. Behav.*, **26**, 293–300.
Schallert, T. & Whishaw, I. Q. (1978). Two types of aphagia and two types of sensorimotor impairment after lateral hypothalamic lesions: observations in normal weight, dieted, and fattened rats. *J. Comp. Physiol. Psychol.*, **92**, 720–41.
Schemmel, R., Michelsen, O. & Mostosky, U. (1970). Influence of body weight, age, diet and sex on fat depots in rats. *Anat. Rec.*, **166**, 437–46.
Schmidt, P. & Andik, I. (1969). Regulation of food intake in parabiotic rats. *Acta Physiol. Acad. Sci. Hung.*, **36**, 293–8.
Schnatz, J. D., Frohman, L. A. & Bernardis, L. L. (1973). The effect of lateral hypothalamic lesions in weanling rats bearing lesions in the ventromedial hypothalamic nuclei. *Proc. Soc. Exp. Biol. Med.*, **142**, 256–7.
Schulkin, J. (1982). Behavior of sodium-deficient rats: the search for the salty taste. *J. Comp. Physiol. Psychol.*, **96**, 628–34.
Schuster, C. R. & Johanson, C. E. (1981). An analysis of drug-seeking behavior in animals. *Neurosci. Biobehav. Rev.*, **5**, 315–23.
Sclafani, A. & Berner, C. N. (1977). Hyperphagia and obesity produced by parasagittal and coronal hypothalamic knife-cuts: further evidence for a longitudinal feeding inhibitory pathway. *J. Comp. Physiol. Psychol.*, **91**, 1000–18.
Sclafani, A., Gale, S. K. & Maul, G. (1974). The effects of knife-cuts between the medial and lateral hypothalamus on feeding and LH self-stimulation in the rat. *Behav. Biol.*, **12**, 491–500.

Sclafani, A. & Maul, G. (1974). Does the ventromedial hypothalamus inhibit the lateral hypothalamus? *Physiol. Behav.*, **12**, 157–62.
Sclafani, A. & Springer, D. (1976). Dietary obesity in adult rats: similarities to hypothalamic and human obesity syndromes. *Physiol. Behav.*, **17**, 461–71.
Scott, E. M. (1946). Self-selection of diet. Appetite for protein. *J. Nutr.*, **32**, 293–301.
(1948). Self selection of diet. *Trans. Am. Assoc. Cereal Chemists*, **6**, 126–33.
Scott, E. M. & Quint, E. (1946). Self selection of diet. III. Appetites for B vitamins. *J. Nutr.*, **32**, 285–91.
Scott, E. M. & Verney, E. L. (1947). Self-selection of diet. VI. The nature of appetites for B vitamins. *J. Nutr.*, **34**, 471–80.
Scott, E. M., Verney, E. L. & Morrissey, P. D. (1950). Self-selection of diet. XI. Appetites for calcium, magnesium and potassium. *J. Nutr.*, **41**, 187–201.
Sensi, S. & Capani, F. (1976). Circadian rhythm of insulin-induced hypoglycemia in man. *J. clin. Endocrinol. Metab.*, **43**, 462–5.
Sensi, S., Capani, F., Caradonna, P. & Dell'acqua, G. B. (1973). Circadian rhythm of immunoreactive insulin under glycemic stimulus. *J. Interdiscip. Cycle Res.*, **4**, 1–14.
Seward, J. P. & Greathouse, S. R. (1973). Appetitive and aversive conditioning in thiamine-deficient rats. *J. Comp. Physiol. Psychol.*, **83**, 157–68.
Seydoux, J., Rohner-Jeanrenaud, F., Assimacopoulos-Jeannet, F., Jeanrenaud, B. & Girardier, L. (1981). Functional disconnection of brown adipose tissue in hypothalamic obesity in rats. *Pflügers Arch.*, **390**, 1–4.
Share, I., Martyniuk, E. & Grossman, M. I. (1952). Effect of prolonged intragastric feeding on oral food intake in dogs. *Am. J. Physiol.*, **169**, 229–35.
Shimazu, T. (1981). Central nervous system regulation of liver and adipose tissue metabolism. *Diabetologia*, **20** (Suppl.), 343–6.
Sims, E. H. A., Danforth, E., Horton, E. S., Bray, G. A., Glennon, J. A. & Salans, L. B. (1973). Endocrine and metabolic effects of experimental obesity in man. *Recent Prog. Horm. Res.*, **29**, 457–96.
Simson, P. C. & Booth, D. A. (1973*a*). Olfactory conditioning by association with histidine-free or balanced amino-acid loads in rats. *J. Exp. Psychol.*, **25**, 354–9.
(1973*b*). Subcutaneous release of amino-acid loads on food and water intakes in the rat. *Physiol. Behav.*, **11**, 329–36.
(1974*a*). Dietary aversion established by a deficient load: specificity to the amino-acid omitted from a balanced mixture. *Pharmac. Biochem. Behav.*, **2**, 481–6.
(1974*b*). The rejection of a diet which has been associated with a single administration of an histidine-free amino-acid mixture. *Br. J. Nutr.*, **31**, 285–96.
Skinner, B. F. (1930). On the conditions of eliciting some eating reflexes. *Proc. natn. Acad. Sci. USA*, **36**, 433–8.
Slonaker, J. R. (1925). The effect of copulation, pregnancy, pseudopregnancy and lactation on the voluntary activity and food consumption of the albino rat. *Am. J. Physiol.*, **71**, 362–94.
Smith, A. L., Satterthwaite, H. S. & Sokoloff, L. (1969). Induction of brain D(−)-β-hydroxybutyrate dehydrogenase activity by fasting. *Science*, **163**, 79–81.
Smith, A. M. (1958). Stimulation of the lateral and medial hypothalamus in rats. *Anat. Rec.*, **104**, 353.
Smith, C. J. V. (1972). Hypothalamic glucoreceptors – the influence of gold thioglucose implants in the ventromedial and lateral hypothalamic areas of normal and diabetic rats. *Physiol. Behav.*, **9**, 391–6.

Smith, C. J. V. & Britt, D. L. (1971). Obesity in the rat induced by hypothalamic implants of gold thioglucose. *Physiol. Behav.*, **7**, 7–10.
Smith, G. P. & Gibbs, J. (1975). Cholecystokinin: a putative satiety signal. *Pharmacol. Biochem. Behav.*, **3** (Suppl. 1), 135–8.
Smith, G. P. & Epstein, A. N. (1969). Increased feeding in response to decreased glucose utilization in the rat and monkey. *Am. J. Physiol.*, **217**, 1083–7.
Smotherman, W. P. (1982). Odor aversion learning by the rat fetus. *Physiol. Behav.*, **29**, 769–72.
Smutz, E. R., Hirsch, E. & Jacobs, H. (1975). Caloric compensation in hypothalamic obese rats. *Physiol. Behav.*, **14**, 305–9.
Snowdon, C. T. (1969). Motivation, regulation and the control of meal parameters with oral and intragastric feeding. *J. Comp. Physiol. Psychol.*, **69**, 91–100.
Snowdon, C. T. & Epstein, A. N. (1970). Oral and intragastric feeding in vagotomized rats. *J. Comp. Physiol. Psychol.*, **71**, 59–67.
Solomon, J., Bulkley, R. J. & Mayer, J. (1974). Effect of a low dose of alloxan on blood glucose, islet beta cell granulation, body weight, and insulin resistance of *ob/ob* mice. *Diabetologia*, **10**, 709–16.
Specchia, G., Fratino, P., Tavazzi, L. & Petroboni, V. (1967). Variazione dell'assimilazione glucidica durante la giornata. *Boll. Soc. Ital. Biol. Sper.*, **43**, 1891–5.
Spiegel, T. A. (1973). Caloric intake of food intake in man. *J. Comp. Physiol. Psychol.*, **84**, 24–37.
Spiegel, T. A. & Jordan, H. A. (1978). Effects of simultaneous oral intragastric ingestion on meal patterns and satiety in humans. *J. Comp. Physiol. Psychol.*, **92**, 133–41.
Spies, G. (1965). Food versus intracranial self-stimulation reinforcement in food-deprived rats. *J. Comp. Physiol. Psychol.*, **60**, 153–7.
Stark, K. A. (1963). Effects of early and prolonged experience with bitter water on its preferableness to guinea pigs. *Dissert. Abstr.*, **24**, 859–60.
Steffens, A. B. (1969*a*). A method for frequent sampling of blood and continuous infusion of fluids in the rat without disturbing the animal. *Physiol. Behav.*, **4**, 833–6.
(1969*b*). The influence of insulin injections and infusions on eating and blood glucose level in the rat. *Physiol. Behav.*, **4**, 823–8.
(1969*c*). Blood glucose and FFA levels in relation to the meal pattern in the normal rat and the ventromedial hypothalamic lesioned rat. *Physiol. Behav.*, **4**, 215–25.
(1970). Plasma insulin content in relation to blood glucose level and meal pattern in the normal and hypothalamic hyperphagic rats. *Physiol. Behav.*, **5**, 147–52.
(1975). Influence of reversible obesity on eating behavior, blood glucose and insulin in the rat. *Am. J. Physiol.*, **228**, 1738–44.
Steffens, A. B. & Lotter, E. C. (1979). Paper given at the International Diabetes Congress, Vienna.
Steinbaum, E. A. & Miller, N. E. (1965). Obesity from eating elicited by daily stimulation of hypothalamus. *Am. J. Physiol.*, **208**, 1–5.
Steiner, J. E. (1973). The gusto-facial response: observation on normal and anencephalic newborn infants. In *Oral Sensation and Perception: Development in the Fetus and Infant*, ed. J. F. Bosma, pp. 254–78. US Dept of Health, Education & Welfare: Bethesda, Maryland.
Stellar, E. & Hill, J. H. (1952). The rat's rate of drinking as a function of water deprivation. *J. Comp. Physiol. Psychol.*, **45**, 96–102.

Stephens, D. N. (1981). Weight gain of young rats fed on a cafeteria diet following excision of interscapular brown adipose tissue. *Proc. Nutr. Soc.*, **40**, A54.

Stevenson, J. A. F., Box, B. M., Feleki, V. & Beaton, J. R. (1966). Bouts of exercise and food intake in the rat. *J. Appl. Physiol.*, **21**, 118–22.

Stickrod, G., Kimble, D. P. & Smotherman, W. P. (1982). *In utero* taste/odor aversion conditioning in the rat. *Physiol. Behav.*, **28**, 5–8.

Stoloff, M. L., Kenny, E. M., Blass, E. M. & Hall, W. G. (1980). The role of experience in suckling maintenance in albino rats. *J. Comp. Physiol. Psychol.*, **94**, 847–56.

Stricker, E. M. & Rowland, N. (1978). Hepatic versus cerebral origin of stimulus for feeding induced by 2-deoxy-D-glucose in rats. *J. Comp. Physiol. Psychol.*, **92**, 126–32.

Stricker, E. M., Swerdloff, A. F. & Zigmond, M. J. (1978). Intrahypothalamic injections of kainic acid produce feeding and drinking deficits in rats. *Brain Res.*, **158**, 470–3.

Strohmayer, A. J., Silverman, G. & Grinker, J. A. (1980). A device for the continuous recording of solid food ingestion. *Physiol. Behav.*, **24**, 789–91.

Strubbe, J. H. & Steffens, A. B. (1977). Blood glucose levels in portal and peripheral circulation and their relation to food intake in the rat. *Physiol. Behav.*, **19**, 303–8.

Strubbe, J. H., Steffens, A. B. & De Ruiter, L. (1977). Plasma insulin and the time pattern of feeding in the rat. *Physiol. Behav.*, **18**, 81–6.

Stunkard, A. J. & Wolff, H. G. (1954). Correlation of arteriovenous glucose differences, gastric hunger contractions and the experience of hunger in man. *Fedn. Proc.*, **13**, 147.

Swerdlow, N. R., Van Der Kooy, D., Koob, G. F. & Wenger, J. R. (1983). Cholecystokinin produces conditioned place-aversions, not place-preferences, in food-deprived rats: evidence against involvement in satiety. *Life Sci.*, **32**, 2087–94.

Symparian, F. M. & MacLendon, P. (1945). Further record of self-demand schedule in infant feeding. *J. Pediat.*, **17**, 109–14.

Tanabe, T., Yarita, H., Iino, M., Ooshima, Y. & Takagi, S. F. (1975). An olfactory projection area in the orbitofrontal cortex of the monkey. *J. Neurophysiol.*, **38**, 1269–81.

Teicher, M. H. & Blass, E. M. (1977). First suckling response of the newborn albino rat: the roles of olfaction and amniotic fluid. *Science*, **198**, 635–7.

Teicher, M. H., Flaum, L. E., Williams, M., Eckhert, S. J. & Lumia, A. R. (1978). Survival, growth and suckling behavior of neonatally bulbectomized rats. *Physiol. Behav.*, **21**, 553–62.

Teitelbaum, P. & Campbell, B. A. (1956). Ingestion patterns in hyperphagic and normal rats. *J. Comp. Physiol. Psychol.*, **51**, 135–41.

Teitelbaum, P., Cheng, M. F. & Rozin, P. (1969). Stages of recovery and development of lateral hypothalamic control of food and water intake. *Ann. N.Y. Acad. Sci.*, **157** (Art. 2), 849–60.

Teitelbaum, P. & Cytawa, J. (1965). Spreading depression and recovery from lateral hypothalamic damage. *Science*, **147**, 61–3.

Teitelbaum, P. & Epstein, A. N. (1962). The lateral hypothalamic syndrome: recovery of feeding and drinking after lateral hypothalamic lesions. *Psychol. Rev.*, **69**, 74–90.

Teitelbaum, P. & Stellar, E. (1954). Recovery from the failure to eat produced by hypothalamic lesions. *Science*, **120**, 894–5.

Tenen, S. S. & Miller, N. E. (1964). Strength of electrical stimulation of lateral

hypothalamus food deprivation and tolerance for quinine in food. *J. Comp. Physiol. Psychol.*, **58**, 55–62.
Thomas, D. W. & Mayer, J. (1968). Meal taking and regulation of food intake by normal and hypothalamic hyperphagic rats. *J. Comp. Physiol. Psychol.*, **66**, 642–53.
(1978). Meal size as a determinant of food intake in normal and hypothalamic obese rats. *Physiol. Behav.*, **21**, 113–18.
Thomas, B. M. & Miller, A. T. (1958). Adaptation to forced exercise in the rat. *Am. J. Physiol.*, **193**, 350–4.
Thompson, D. A. & Campbell, R. G. (1978). Experimental hunger in man: behavioral and metabolic correlates of intracellular glucopenia. In *Central Mechanisms of Anorectic Drugs*, ed. S. Garattini & R. Samanin, pp. 437–45. Raven Press: New York.
Treit, D., Spetch, M. L. & Deutsch, J. A. (1983). Variety in the flavor of food enhances eating in the rat: A controlled demonstration. *Physiol. Behav.*, **30**, 207–12.
Turro, R. (1914). *Les Origines de la Connaissance*, F. Alcan: Paris.
Ungerstedt, U. (1971). Adipsia and aphagia after 6-hydroxydopamine induced degeneration of the nigrostriatal dopamine system. *Acta Physiol. Scand. Suppl.*, **367**, 95–122.
Usami, M., Seino, Y., Seino, S., Takemura, J., Nakahara, H., Ikeda, M. & Imura, H. (1982). Effects of high-protein diet on insulin and glucagon secretion in normal rats. *J. Nutr.*, **112**, 681–5.
Valenstein, E. S. & Cox, V. C. (1970). Influence of hunger, thirst and previous experience in the test chamber on stimulus-bound eating and drinking. *J. Comp. Physiol. Psychol.*, **70**, 189–99.
Valenstein, E. S., Cox, V. C. & Kakolewski, J. W. (1968). Modification of motivated behavior elicited by electrical stimulation of the hypothalamus. *Science*, **159**, 1119–21.
Valle, F. P. (1968). Effect of exposure to feeding-related stimuli on food consumption in rats. *J. Comp. Physiol. Psychol.*, **66**, 773–6.
Van der Tuig, J. G., Knehans, A. W. & Romsos, D. R. (1982). Reduced sympathetic nervous system activity in rats with ventromedial hypothalamic lesions. *Life Sci.*, **30**, 913–20.
Van Houten, M. & Posner, B. I. (1981). Insulin receptors in the central nervous system: localization and characteristics. In *Current Views and Insulin Receptors*, ed. D. Andreani, R. de Pirro, R. Lauro, J. M. Olefski & J. Roth, pp. 75–90. Academic Press: New York.
Van Vort, W. & Smith, G. P. (1983). The relationships between the positive reinforcing and satiating effects of a meal in the rat. *Physiol. Behav.*, **30**, 279–84.
Varner, L. H. (1928). The study of hunger behavior in the rat by means of the obstruction method. *J. Comp. Physiol. Psychol.*, **8**, 273–300.
Wakerley, J. B. & Drewett, R. F. (1975). Pattern of suckling in the infant rat during spontaneous milk ejection. *Physiol. Behav.*, **15**, 277–82.
Walike, B. C. & Smith, O. A. (1972). Regulation of food intake during intermittent and continuous cross circulation in monkeys (*Macaca mulatta*). *J. Comp. Physiol. Psychol.*, **80**, 372–81.
Walls, E. V. & Wishart, T. B. (1977). Reliable method for cannulation of the third ventricle of the rat. *Physiol. Behav.*, **19**, 171–3.

Wang, G. H. (1923). Relation between 'spontaneous' activity and oestrus cycle in the white rat. *Comp. Psychol. Mongr.*, **2**, 1–27.
Warner, K. E. & Balagura, S. (1975). Intrameal eating patterns of obese and non-obese humans. *J. Comp. Physiol. Psychol.*, **89**, 778–83.
Watson, P. J., Short, M. A. & Hartman, D. F. (1979). Re-emergence of hypothalamically elicited eating following change in food. *Physiol. Behav.*, **23**, 663–8.
Wayner, M. J., Cott, A., Millner, J. & Tartaglione, R. (1971). Loss of 2-deoxy-D-glucose induced eating in recovered lateral rats. *Physiol. Behav.*, **7**, 881–4.
Wayner, M. J., Ono, T., de Young, A. & Barone, F. C. (1975). Effects of essential amino acids on central neurons. *Pharmac. Biochem. Behav.*, **3** (Suppl. 1), 85–90.
Wayner, M. J., Yin, T. H., Barone, F. C., Lee, H. K. & Tsai, C. T. (1979). Effects of discrete destruction of functionally identified chemosensitive hypothalamic neurons of ingestive behavior. *Physiol. Behav.*, **23**, 385–90.
Weingarten, H. P., Chang, P. & Jarvie, K. R. (1983). Reactivity of normal and VMH-lesion rats to quinine adulterated foods: negative evidence for negative finickiness. *Behav. Neurosci.*, **97**, 221–33.
Widdowson, E. M. (1962). Nutritional individuality. *Proc. Nutr. Soc.*, **21**, 121–8.
Wiepkema, P. R. (1971). Positive feedbacks at work during feeding. *Behaviour*, **39**, 266–73.
Wiepkema, P. R., De Ruiter, L. & Reddingius, J. (1966). Circadian rhythms in the feeding behaviour of CBA mice. *Nature*, **209**, 935–6.
Wilkinson, H. A. & Peele, T. L. (1962). Modification of intracranial self-stimulation by hunger satiety. *Am. J. Physiol.*, **203**, 537–40.
Williams, R. A. & Campbell, B. A. (1961). Weight loss and quinine-milk ingestion as measures of 'hunger' in infant and adult rats. *J. Comp. Physiol. Psychol.*, **54**, 220–2.
Wirth, J. B. & McHugh, P. R. (1983). Gastric distension and short-term satiety in the rhesus monkey. *Am. J. Physiol.*, **245**, R174–R180.
Wise, R. A. (1968). Hypothalamic motivational systems: fixed or plastic neural circuits. *Science*, **162**, 377–9.
Wise, R. A., Spindler, J., de Wit, H. & Gerber, G. J. (1978). Neuroleptics-induced 'anhedonia' in rats: pimozide blocks reward quality of food. *Science*, **201**, 262–4.
Wolf, L. L. & Hainsworth, F. R. (1977). Temporal patterning of feeding by humming birds. *Anim. Behav.*, **25**, 976–89.
Woods, S. C. (1976). Conditioned hypoglycemia. *J. Comp. Physiol. Psychol.*, **90**, 1164–8.
Woods, S. C., Lotter, E. C., McKay, L. D. & Porte, D., Jr (1979). Chronic intracerebroventricular infusion of insulin reduces food intake and body weight of baboons. *Nature*, **282**, 503–5.
Woods, S. C. & McKay, L. D. (1978). Intraventricular alloxan eliminates feeding elicited by 2-deoxy-glucose. *Science*, **202**, 1209–10.
Woods, S. C., Vasselli, J. R., Kaestner, E., Szakmari, G. A., Milburn, P. & Vitiello, M. V. (1977). Conditioned insulin secretion and meal feeding in rats. *J. Comp. Physiol. Psychol.*, **91**, 128–33.
Wooley, O. W., Wooley, S. C. & Dunham, R. B. (1972). Can calories be perceived and do they affect hunger in obese and nonobese humans? *J. Comp. Physiol. Psychol.*, **80**, 250–8.

Wurtman, R. J. & Wurtman, J. J. (1977). *Nutrition and the Brain*, vol. 1, *Determinants of the Availability of Nutrients to the Brain*. Raven Press: New York.
Wurtman, J. J. & Wurtman, R. J. (1979). Drugs that enhance central serotoninergic transmission diminish elective carbohydrate consumption by rats. *Life Sci.*, **24**, 895–904.
Yamamoto, T. & Shibata, Y. (1979). Direct fiber connexion between frontal cortex and the hypothalamus in rats. *Pharmac. Biochem. Behav.*, 3 (Suppl.), 15–22.
Yokel, R. A. & Wise, R. A. (1975). Increased lever pressing for amphetamine after pimozide: implications for a dopamine theory of reward. *Science*, **187**, 547–9.

Index